SpringerBriefs in Mathematics

SpringerBriefs in Mathematics showcases expositions in all areas of mathematics and applied mathematics. Manuscripts presenting new results or a single new result in a classical field, new field, or an emerging topic, applications, or bridges between new results and already published works, are encouraged. The series is intended for mathematicians and applied mathematicians.

More information about this series at http://www.springer.com/series/10030

SpringerBriefs present concise summaries of cutting-edge research and practical applications across a wide spectrum of fields. Featuring compact volumes of 50 to 125 pages, the series covers a range of content from professional to academic. Briefs are characterized by fast, global electronic dissemination, standard publishing contracts, standardized manuscript preparation and formatting guidelines, and expedited production schedules.

Typical topics might include:

- A timely report of state-of-the art techniques
- A bridge between new research results, as published in journal articles, and a contextual literature review
- A snapshot of a hot or emerging topic
- An in-depth case study
- A presentation of core concepts that students must understand in order to make independent contributions

Titles from this series are indexed by Web of Science, Mathematical Reviews, and zbMATH.

John Toland

The Dual of $L_\infty(X, \mathcal{L}, \lambda)$, Finitely Additive Measures and Weak Convergence

A Primer

 Springer

John Toland
Department of Mathematical Sciences
University of Bath
Bath, UK

ISSN 2191-8198   ISSN 2191-8201   (electronic)
SpringerBriefs in Mathematics
ISBN 978-3-030-34731-4   ISBN 978-3-030-34732-1   (eBook)
https://doi.org/10.1007/978-3-030-34732-1

Mathematics Subject Classification (2010): 46E30, 28C15, 46T99, 26A39, 28A25, 46B04

This Springer imprint is published by the registered company Springer Nature Switzerland AG
The registered company address is: Gewerbestrasse 11, 6330 Cham, Switzerland

Preface

Assuming some familiarity with Lebesgue measure, integration and related functional analysis summarised in Chap. 2, this is an exposition of topics that arise when identifying elements of the dual space of $L_\infty(X, \mathcal{L}, \lambda)$ with finitely additive measures on a σ-algebra $\mathcal{L}$ when the measure λ is complete and σ-finite. Such a representation has its origins in the independent work of Fichtenholz and Kantorovitch [14] and Hildebrandt [20] and culminated in a general abstract theory due to Yosida and Hewitt [35] in 1952. However, even now it is not unusual ([16] is an exception) for books on measure theory to give a detailed account of the dual space of $L_p(X, \mathcal{L}, \lambda)$, $1 \leqslant p < \infty$, while relegating the case $p = \infty$ to references, e.g. [12, p. 296] which invokes the theory of finitely additive measures on algebras such as is developed in [35].

An explanation may be that a study of finitely additive measures on algebras necessitates a possibly unwelcome diversion from the mainstream theory of countably additive measures that suffices for $p \in [1, \infty)$. Whatever the reason, a consequence is that $L_\infty(X, \mathcal{L}, \lambda)^*$ has acquired an aura of mystery, to the extent that it is often not very clear beyond the mere definition what is meant by saying that a bounded sequence in $L_\infty(X, \mathcal{L}, \lambda)$ is weakly convergent.

The aim here is to take Yosida and Hewitt theory on σ-algebras beyond the representation theorem for $L_\infty(X, \mathcal{L}, \lambda)^*$, pointing out some of its consequences for measurable functions generally and in particular for weak convergence of sequences in $L_\infty(X, \mathcal{L}, \lambda)$. The target audience is anyone who feels nervous about representing the dual of $L_\infty(X, \mathcal{L}, \lambda)$ by finitely additive measures in the knowledge that *there exist uncountably many, linearly independent, finitely additive measures $v \geqslant 0$ defined on the Lebesgue σ-algebra of $(0, 1)$ with the property that*

$$\int_0^{1 - \frac{1}{k}} u \, dv = 0 \text{ for all } u \in L_\infty(0, 1) \text{ and } k \in \mathbb{N}, \text{ but } \int_0^1 1 \, dv = 1. \qquad (\dagger)$$

An essential goal will be to come to terms with observations such as this one.

In their seminal work, Yosida and Hewitt [35] studied general Banach spaces $L_\infty(X,\mathcal{M},\mathcal{N})$ of essentially bounded measurable functions, where measurability is determined by an algebra $\mathcal{M}$ (closed under complementation and finite unions) and essential boundedness is defined in terms of a family $\mathcal{N} \subset \mathcal{M}$ (closed under countable unions with the added property that $A \subset B \in \mathcal{N}$ implies $A \in \mathcal{N}$) that mimics null sets. Obviously, $L_\infty(X,\mathcal{L},\lambda)$ is a special case of $L_\infty(X,\mathcal{M},\mathcal{N})$ but in general no measure of any kind is involved in the definition of $L_\infty(X,\mathcal{M},\mathcal{N})$. However, although [35] shows that the dual of $L_\infty(X,\mathcal{M},\mathcal{N})$ can be expressed in terms of finitely additive measures, the exposition here is restricted to $L_\infty(X,\mathcal{L},\lambda)$ because

properties of finitely additive measures on σ-algebras are less circumscribed by hypotheses than on algebras, and replacing the algebra $\mathcal{M}$ by a σ-algebra $\mathcal{L}$ and $\mathcal{N}$ by the family of null sets $\{E \in \mathcal{L} : \lambda(E) = 0\}$, where λ is complete and σ-finite, yields a theory which is relevant in applications, including when X is a Lebesgue measurable subset of $\mathbb{R}^n$ or a differentiable manifold, or when $X = \mathbb{N}$ with counting measure.

For a σ-finite measure space the ultimate aim is to develop theory sufficient to characterise weakly convergent sequences in $L_\infty(X,\mathcal{L},\lambda)$ in terms of their λ-almost-everywhere pointwise behaviour. However, in the process, when (X,ρ) is a locally compact Hausdorff topological space and $(X,\mathcal{B},\lambda)$ is a corresponding Borel measure space, there emerges a natural way to localise weak convergence. A sequence is weakly convergent in $L_\infty(X,\mathcal{B},\lambda)$ if and only if it is weakly convergent at every point x_0 in the one-point compactification of (X,ρ). Here, weak convergence at x_0 is defined in terms of functionals which are zero outside every neighbourhood of x_0; for an example of such, see (†).

The essential range $\mathcal{R}(u)(x_0)$[1] of a Borel measurable function u at x_0 is similarly defined in terms of those elements of $L_\infty(X,\mathcal{B},\lambda)^*$ which are localised at x_0. Since it need not be a singleton, $\mathcal{R}(u)(x_0)$ can be interpreted as a multivalued representation of the fine structure at x_0 of $u \in L_\infty(X,\mathcal{B},\lambda)$ which is intimately related to weak convergence.

The Literature

In her Foreword to the monograph by Bhaskara Rao and Bhaskara Rao [6], Dorothy Maharam Stone cites Salomon Bochner as having said that "contrary to popular mathematical opinion finitely additive measures were more interesting, more difficult to handle, and perhaps more important than countably additive measures". Oxtoby [25] described [6] as a comprehensive account of finitely additive measures which effectively organises a large body of material that is widely scattered in the literature and deserves to be better known, and in their preface the authors themselves described it as a reference book as well as a textbook.

---

[1] $\mathcal{R}(u)(x_0)$ is sometimes referred to as the cluster set of u at x_0.

The origins of this theory are to be found in the early days of modern integration theory when there were many contributors: see [12, §III.15, p. 233 and §IV.16, p. 388[2]] and the comprehensive bibliography with notes in [6]. However, presumably because they could not match the versatility of Lebesgue's theory of integration and the power of its convergence theorems, finitely additive measures seem to have fallen out of fashion. Nevertheless, they continue to have significant roles in, for example, mathematical economics, probability, statistics, optimization, control theory and analysis [7, 9, 25, 35].

In a series of three papers on additive set functions on abstract topological spaces, A. D. Alexandroff [2] studied bounded regular finitely additive measures that represent linear functionals on spaces of continuous functions. On a similar theme, but in a more general setting, a much-cited reference for the dual of $L_\infty(X, \mathcal{L}, \lambda)$ is Dunford and Schwartz [12, p. 296] which covers the theory of finitely additive set functions on algebras and includes extensive historical notes. For a recent account, see Fonseca and Leoni [16, Theorem 2.24], and Aliprantis and Border [3] for the abstract theory in which it is embedded.

It will soon be apparent that key results, including (†), rely on the axiom of choice. For a discussion of the role of the axiom of choice, geometrical and paradoxical aspects of finitely additive measures, and their invariance under group actions on X, see Tao [32]. Oxtoby's commentary [25] is of independent interest.

A key role is played throughout by the set $\mathfrak{G}$ of finitely additive measures that take only the values 0 and 1 and the observation that $L_\infty(X, \mathcal{L}, \lambda)$ is isometrically isomorphic to a space of real-valued continuous functions on $(\mathfrak{G}, \tau)$ with the maximum norm, where τ is a compact Hausdorff topology. Further analysis of $\mathfrak{G}$ in a Borel setting then leads to localization, and to other developments mentioned above and outlined in the Introduction.

What follows is in large part an extension of a simplified version of Yosida and Hewitt [35], set out in the notation and terminology of Chap. 2.

Bath, UK   John Toland

Acknowledgements I am indebted to Charles Stuart (Lausanne) for many things including his encouragement of this project. I am grateful to Anthony Wickstead (Belfast) who obtained for me a copy of [33] and drew my attention to [34], and to Geoffrey Burton (Bath) and Eugene Shargorodsky (King's London) for their interest and many comments. In addition, I would like to thank Eugene Shargorodsky who contributed Sect. 9.4 and Mauricio Fernández (Stuttgart) who on a visit to Cambridge asked a question that the account that follows attempts to answer.

---

[2]The reference to Theorem 8.15 on p. 388 is a misprint; 8.15 is a Definition and obviously Theorem 8.16 was intended.

Contents

1 **Introduction** 1

2 **Notation and Preliminaries** 7
   2.1 Partial Ordering and Zorn's Lemma 7
   2.2 Algebras and σ-Algebras 8
   2.3 Measurable Sets and Measurable Functions 9
   2.4 Measures and Real Measures 10
   2.5 Splitting Families of Measurable Sets 14
   2.6 Integration 15
   2.7 Function Spaces 17
   2.8 Dual Spaces and Measures 19
   2.9 Functional Analysis 20
   2.10 Point Set Topology 23

3 L_∞ **and Its Dual** 27

4 **Finitely Additive Measures** 31
   4.1 Definition, Notation and Basic Properties 31
   4.2 Purely Finitely Additive Measures 36
   4.3 Canonical Decomposition: $\mathrm{ba}(\mathcal{L}) = \Sigma(\mathcal{L}) \oplus \Pi(\mathcal{L})$ 38
   4.4 $L_\infty^*(X, \mathcal{L}, \lambda)$ 39

5 $\mathfrak{G}$**: 0–1 Finitely Additive Measures** 41
   5.1 $\mathfrak{G}$ and Ultrafilters 42
   5.2 $\mathfrak{G}$ and the λ-Finite Intersection Property 44

6   Integration and Finitely Additive Measures 47
   6.1   The Integral 47
   6.2   Yosida–Hewitt Representation: Proof of Theorem 3.1 50
   6.3   Integration with Respect to $\omega \in \mathfrak{G}$ 51
   6.4   Essential Range of $u \in L_\infty(X, \mathcal{L}, \lambda)$ 52
   6.5   Integrating $u \in \ell_\infty(\mathbb{N})$ with Respect to $\mathfrak{G}$ 52
   6.6   The Valadier–Hensgen Example 54

7   Topology on $\mathfrak{G}$ 57
   7.1   The Space $(\mathfrak{G}, \tau)$ 57
   7.2 $L_\infty(X, \mathcal{L}, \lambda)$ and $C(\mathfrak{G}, \tau)$ Isometrically Isomorphic 58
   7.3   Properties of $\mathfrak{G}$ and τ 59
   7.4 $\mathfrak{G}$ and the Weak* Topology on $L_\infty^*(X, \mathcal{L}, \lambda)$ 63
   7.5 $\mathfrak{G}$ as Extreme Points................................ 64

8   Weak Convergence in L_∞ $(X, \mathcal{L}, \lambda)$ 67
   8.1   Weakly Convergent Sequences 67
   8.2   Pointwise Characterisation............................. 68
   8.3   Applications of Theorem 8.7 72

9 L_∞^* When X is a Topological Space 77
   9.1   Localising $\mathfrak{G}$ 77
   9.2   Localising Weak Convergence.......................... 80
   9.3   Fine Structure at x_0 of $u \in L_\infty(X, \mathcal{B}, \lambda)$ 81
   9.4   A Localised Range from Complex Function Theory 84

10   Reconciling Representations 87
   10.1 $(\mathfrak{G}, \tau)$ Versus $L_\infty(X, \mathcal{L}, \lambda)$ 87
   10.2   Restriction to $C_0(X, \varrho)$ of Elements of $L_\infty^*(X, \mathcal{B}, \lambda)$ 89

References 95

Index 97

Chapter 1
Introduction

Overview

In a normed linear space V, a sequence $\{v_k\}$ converges weakly to v ($v_k \rightharpoonup v$) if $v^*(v_k) \to v^*(v)$ for all $v^* \in V^*$, the dual space of V and, from the uniform boundedness principle, weakly convergent sequences are bounded in norm. However, it has been known since the work of Banach that when V is a complete normed linear space it may not be necessary to use all elements of V^* when testing for weak convergence. Indeed, when $C(Z)$ denotes the space of real-valued continuous functions on a compact metric space Z with the maximum norm, he showed that $v_k \rightharpoonup v$ in $C(Z)$ if and only if $v_k(z) \to v(z)$ for all $z \in Z$ and $\{\|v_k\|\}$ is bounded. To do so he observed [5, Annexe, Thm. 7] that Dirac δ-functions satisfy conditions for a set W^* in the dual space of a Banach space to have the property that

$$\{\|v_k\|\} \text{ bounded and } w^*(v_k) \to 0 \text{ for all } w^* \in W^* \text{ imply } v_k \rightharpoonup 0. \qquad \text{(W)}$$

When (X, ϱ) is a locally compact Hausdorff space and $(C_0(X, \varrho), \|\cdot\|_\infty)$ is the Banach space of real-valued continuous functions on X that vanish at infinity (see (2.9)), weakly convergent sequences are pointwise convergent because δ-functions belongs to the dual space of $C_0(X, \varrho)$, and bounded by the uniform boundedness principle. Conversely, from Theorem 2.37 (Riesz) and Lebesgue's Dominated Convergence Theorem [15, Thm. 2.24], sequences that are norm-bounded and pointwise convergent on X are weakly convergent.

In particular, when $\mathcal{Z}$ is a compact Hausdorff space, for $\{v_k\} \subset C(\mathcal{Z})$ (the space of real-valued continuous functions on $\mathcal{Z}$ with the maximum norm)

$$v_k \rightharpoonup v_0 \text{ in } C(\mathcal{Z}) \Leftrightarrow \sup_k \|v_k\| < \infty \text{ and } v_k(z) \to v_0(z) \text{ for all } z \in \mathcal{Z}. \qquad \text{(V)}$$

The possibility of usefully extending these observations to $L_\infty(X, \mathcal{L}, \lambda)$ (the real Banach space of essentially bounded real-valued functions defined by (2.8)) at first appears limited because, for example, in an open set $\Omega \subset \mathbb{R}^n$ with Lebesgue measure,

© The Author(s), under exclusive license to Springer Nature Switzerland AG 2020   1
J. Toland, *The Dual of $L_\infty(X, \mathcal{L}, \lambda)$, Finitely Additive Measures
and Weak Convergence*, SpringerBriefs in Mathematics,
https://doi.org/10.1007/978-3-030-34732-1_1

$u_k \rightharpoonup u$ in $L_\infty(\Omega)$ implies that $u_k(x) \to u(x)$ almost everywhere and $\{\|u_k\|_\infty\}$ is bounded, but the converse is false (Lemma 8.4 and Example 8.6). Nevertheless, for bounded sequences it is shown in Theorem 8.7 that weak convergence is equivalent to strong convergence in $L_\infty(X, \mathcal{L}, \lambda)$ of elements of a family of related sequences and the resultant test is illustrated by several examples in Sect. 8.3.

The proof, which *inter alia* is developed from first principles in the following pages, depends on the construction a compact Hausdorff topological space $(\mathfrak{G}, \tau)$ with the property that $L_\infty(X, \mathcal{L}, \lambda)$ and $C(\mathfrak{G}, \tau)$ are isometrically isomorphic (Theorem 7.4). Here $\mathfrak{G}$ (Gothic G) is the set of the finitely additive measures that take only values 0 and 1 on $\mathcal{L}$, and are 0 on null sets in a general measure space $(X, \mathcal{L}, \lambda)$.

When the essential range of $u \in L_\infty(X, \mathcal{L}, \lambda)$ is defined as

$$\mathcal{R}(u) = \left\{\alpha \in \mathbb{R} : \lambda(\{x : |u(x) - \alpha| < \epsilon\}) > 0 \text{ for all } \epsilon > 0\right\},$$

it is shown that

$$\mathcal{R}(u) = \left\{\int_X u \, d\omega : \omega \in \mathfrak{G}\right\}$$

and its relation to the isometric isomorphism is given by (7.2) and (7.3). If the setting is a Borel measure space $(X, \mathcal{B}, \lambda)$ corresponding to a locally compact Hausdorff spaces (X, ϱ), these considerations can be localised to points x_0 in the one-point compactification X_∞ of X. This in turn leads to the definition of the essential range of $u \in L_\infty(X, \mathcal{B}, \lambda)$ localised at $x_0 \in X_\infty$ and the possibility of regarding u as multivalued at every point while at the same time being single-valued almost everywhere. These ideas are closely related to the dual of $L_\infty(X, \mathcal{B}, \lambda)$ and weak convergence.

The fact that $L_\infty(X, \mathcal{L}, \lambda)$ and $C(\mathfrak{G}, \tau)$ are isometrically isomorphic appears to be at variance with $L_\infty(X, \mathcal{L}, \lambda)^*$ being represented by finitely additive measures while $C(\mathfrak{G}, \tau)^*$ is represented by σ-additive measures. Moreover, when (X, ϱ) is a locally compact Hausdorff space, elements of $L_\infty(X, \mathcal{B}, \lambda)^*$ are represented by finitely additive measures but their restrictions to $C_0(X, \varrho)$ are also represented by Borel measures. Both these issues will be addressed.

As explained in the Preface, finitely additive measures will be considered only on σ-algebras. The classical text [12] and the exhaustive monograph [6] both consider more general situations from the outset, as does Yosida and Hewitt [35] which is the motivation and main source for this account of the simplified theory. The approach here is referred to as Yosida–Hewitt theory.

Layout

Chapter 2 is a brief survey of background material, collected with references from different sources and organised in a consistent notation that will be used in the chapters that follow. Although most of the material is entirely standard and does not need to be digested unless a need arises in later chapters, some of it may be less familiar. For example, Sect. 2.5 is a slight extension of Kirk's [22] application of Baire's category theorem to show that if $(X, \mathcal{L}, \lambda)$ has no atoms and $\mathcal{G}$ is a denumerable family of sets of positive measure, there exists an infinite family of disjoint sets $A \in \mathcal{L}$ such that $\lambda(G \cap A) > 0$ and $\lambda(G \setminus A) > 0$ for all $G \in \mathcal{G}$. This construction is used in Example 9.11 when considering possibilities for the localised essential range of $u \in L_\infty(X, \mathcal{L}, \lambda)$ at x_0 defined in (9.4).

Chapter 3 begins by stating Theorem 3.1, which is the Yosida–Hewitt representation of the dual space $L_\infty(X, \mathcal{L}, \lambda)^*$ as the space of finitely additive measures on $\mathcal{L}$ which are zero on sets of zero λ-measure. These finitely additive measures will be denoted by $L_\infty^*(X, \mathcal{L}, \lambda)$, which should therefore be identified with, but not confused with, $L_\infty(X, \mathcal{L}, \lambda)^*$, the bounded, real-valued, linear functionals on $L_\infty(X, \mathcal{L}, \lambda)$. From Theorem 3.1, it follows intuitively that if $\{A_k\} \subset \mathcal{L}$ is any sequence of mutually disjoint sets, then $\chi_{A_k} \rightharpoonup 0$ in $L_\infty(X, \mathcal{L}, \lambda)$ as $k \to \infty$, where χ_{A_k} denotes the characteristic function of A_k. Moreover, there are finitely additive measures that represent non-zero bounded linear functionals f on $L_\infty(0, 1)$ with $f(v) = 0$ when v is continuous. Since no such countably additive measure can exist because of the dominated convergence theorem and Theorem 2.35 (Lusin), observations such as these reflect the delicacy of Theorem 3.1 and the differences between σ-additive and finitely additive measures.

Chapter 4 introduces basic notation and definitions, for example, of partial ordering, lattice operations, absolute continuity and singularity, for finitely additive measures. The important notion of pure finite additivity of measures follows, and it is shown that every finitely additive measure is uniquely the sum of a σ-additive and a purely finitely additive measure.

Chapter 5 introduces the set $\mathfrak{G}$ of elements in $L_\infty^*(X, \mathcal{L}, \lambda)$ which take only the values 0 or 1 on $\mathcal{L}$ and explains the sense in which every $u \in L_\infty(X, \mathcal{L}, \lambda)$ is constant ω-almost everywhere when $\omega \in \mathfrak{G}$. Elements $\omega \in \mathfrak{G}$ give rise to families of sets $\mathcal{U}_\omega = \{E \in \mathcal{L} : \omega(E) = 1\}$ which are ultrafilters (Definition 5.3). The collection of ultrafilters is denoted by $\mathfrak{U}$ (Gothic U), and there is a one-to-one correspondence (Theorem 5.4) between $\mathfrak{G}$ and $\mathfrak{U}$. The existence of ultrafilters, and consequently of elements of $\mathfrak{G}$ with certain properties, follow from Zorn's lemma. $\mathfrak{G}$ will dominate subsequent developments.

Chapter 6 defines the integral of essentially bounded measurable functions with respect to the finitely additive measures that featured in Theorem 3.1 (the Yosida–Hewitt representation theorem). In Theorem 6.2, it is noted that for all $\omega \in \mathfrak{G}$ and $u \in L_\infty(X, \mathcal{L}, \lambda)$, $u(x) = \int_X u \, d\omega$, ω-almost everywhere in the sense of finitely additive measures (Remark 5.2). Observation (†) in the Preface is justified in Remark 6.1. The chapter ends with an account of the Valadier–Hensgen example [19, 33] of purely finitely additive measures on $[0, 1]$ that are definitely not σ-additive but which coincide with σ-additive Lebesgue measure when integrating continuous functions.

Chapter 7 introduces a compact Hausdorff topology τ on $\mathfrak{G}$ and from theory already developed, derives the existence, Theorem 7.4, of an isometric isomorphism between the Banach algebra $L_\infty(X, \mathcal{L}, \lambda)$ and the space of real-valued continuous functions $C(\mathfrak{G}, \tau)$. It is immediate that the functionals corresponding to $\mathfrak{G}$ have property (W) in the Introduction.

Independently, it is shown that τ coincides with the restriction to $\mathfrak{G}$ of the weak* topology on $L_\infty^*(X, \mathcal{L}, \lambda)$ and that $\mathfrak{G}$ is a closed subset of $L_\infty^*(X, \mathcal{L}, \lambda)$ with the weak* topology. Also, it is shown that $\pm\mathfrak{G}$ coincide with the extreme points of the closed unit ball in $L_\infty^*(X, \mathcal{L}, \lambda)$ and consequently Theorem 2.55 (Rainwater) yields an alternative proof that $\mathfrak{G}$ has property (W) in the Preface.

Chapter 8 opens with the observation, based of the duality between Dirac measures acting on $C(\mathfrak{G}, \tau)$ and elements of $\mathfrak{G}$ acting on $L_\infty(X, \mathcal{L}, \lambda)$, that $u_k \rightharpoonup u_0$ in $L_\infty(X, \mathcal{L}, \lambda)$ if and only if

$$\|u_k\|_\infty \leqslant M \text{ and } \int_X u_k \, d\omega \to \int_X u_0 \, d\omega \text{ as } k \to \infty \text{ for all } \omega \in \mathfrak{G}.$$

It follows that when $F : \mathbb{R} \to \mathbb{R}$ is continuous, $u \mapsto F(u)$ is sequentially weakly continuous [4] on $L_\infty(X, \mathcal{L}, \lambda)$. Necessary pointwise conditions for a sequence to be weakly convergent in $L_\infty(X, \mathcal{L}, \lambda)$ are given but examples show that they are not sufficient. However, a necessary and sufficient pointwise condition for a sequence to be weakly convergent in $L_\infty(X, \mathcal{L}, \lambda)$, Theorem 8.7, follows from Theorem 5.6 and the fact that any $u \in L_\infty(X, \mathcal{L}, \lambda)$ is a constant ω-almost everywhere in the sense of finitely additive measures when $\omega \in \mathfrak{G}$. In Sect. 8.3, some quite subtle questions about the weak convergence of specific sequences are settled using Theorem 8.7.

Chapter 9 deals with refinements of the theory to measure spaces $(X, \mathcal{B}, \lambda)$ where (X, ϱ) is a locally compact Hausdorff space with measure λ on its Borel σ-algebra $\mathcal{B}$. Prototypical examples of this setting are when X is an open subset of $\mathbb{R}^n$ with the Euclidian metric and Lebesgue measure and $X = \mathbb{N}$ with the discrete topology and counting measure. The key observation is that $\mathfrak{G} = \bigcup_{x_0 \in X_\infty} \mathfrak{G}(x_0)$ where, for distinct points $x_0 \in X_\infty$ (the one-point compactification of X), the sets $\mathfrak{G}(x_0)$ are closed in $(\mathfrak{G}, \tau)$ and disjoint, and elements of $\mathfrak{G}(x_0)$ are zero outside every open neighbourhood of $x_0 \in X_\infty$.

In Sect. 9.2, this localisation result leads to a characterization of weakly convergent sequences in terms of the pointwise behaviour of related sequences of functions in neighbourhoods of points of X_∞. (Here "pointwise" has its usual λ-almost everywhere meaning which is familiar in any measure space, whereas "localisation" refers to the behaviour of Borel measures and functions restricted to open neighbourhoods of points in a topological space.)

Writing $\mathcal{R}(u)$, $u \in L_\infty(X, \mathcal{B}, \lambda)$, as a union of disjoint compact sets

$$\mathcal{R}(u) = \bigcup_{x_0 \in X_\infty} \mathcal{R}(u)(x_0) \text{ where } \mathcal{R}(u)(x_0) = \left\{ \int_X u \, d\omega, \omega \in \mathfrak{G}(x_0) \right\}$$

localises the essential range and reflects the fine structure of u at x_0.

As a consequence of Theorem 2.25, if (X, ϱ) is completely separable and $(X, \mathcal{B}, \lambda)$ has no atoms, there exist $u \in L_\infty(X, \mathcal{B}, \lambda)$ such that $u(x) \in \mathbb{Q}$ (the rational numbers) for all $x \in X$ yet $\mathcal{R}(u)(x)$ is the closed interval $[0, \|u\|]$ for all x. The chapter ends with an example due to Shargorodsky of a similar phenomenon occurring naturally in the theory of complex Hardy spaces, see Sect. 9.4.

Chapter 10 first reconciles the general fact that $L_\infty(X, \mathcal{L}, \lambda)$ and $C(\mathfrak{G}, \tau)$ are isometrically isomorphic while the dual of $L_\infty(X, \mathcal{L}, \lambda)$ is represented by finitely additive measures, whereas the dual of $C(\mathfrak{G}, \tau)$ is represented by regular, real Borel measures which are σ-additive. It goes on to consider the special case $L_\infty(X, \mathcal{B}, \lambda)$ where (X, ϱ) is a locally compact Hausdorff space and $\mathcal{B}$ denotes the Borel subsets of X. In that case, elements of $L_\infty(X, \mathcal{B}, \lambda)^*$ are represented by finitely additive measures but when restricted to $C_0(X, \varrho)$ they are also represented by Borel measures.

As seen in Chap. 6, Valadier and Hensgen independently noted that Riemann sums have Banach limits (Definition 2.41) which on $L_\infty[0, 1]$ are represented by purely finitely additive measures which are not σ-additive, but which coincide with the Lebesgue integral for continuous functions. That observation is the motivation for Sect. 10.2 which considers the relation between the finitely additive measures ν that yield elements of $L_\infty(X, \mathcal{B}, \lambda)^*$ (Theorem 3.1) and the Borel measures $\hat{\nu}$ that by Theorem 2.37 (Riesz) represent their restrictions to $C_0(X, \varrho)$. In particular, those ν for which $\hat{\nu}$ is singular with respect to λ are characterised in Corollary 10.10 and a minimax formula for $\hat{\nu}$ in terms of ν is given in Theorem 10.11. It follows that $\hat{\nu}$ may be zero when $\nu \geqslant 0$ if (X, ϱ) is not compact.

When $\omega \in \mathfrak{G}$, either $\hat{\omega} \in \mathfrak{D}$ (a Dirac measures on X) or $\hat{\omega}$ may be zero if (X, ϱ) is not compact; if (X, ϱ) is compact $\hat{\omega} \in \mathfrak{D}$. Note from Remark 6.7 that an arbitrary Hahn–Banach extension to $L_\infty(X, \mathcal{B}, \lambda)$ of a Dirac δ-function acting on $C_0(X, \varrho)$ need not be in $\mathfrak{G}$ and from Chap. 9 there may be infinitely many extensions that are in $\mathfrak{G}$.

Starting from (and referring to Chap. 2 only when necessary), a self-contained approach to the pointwise characterisation of weakly convergent sequences is presented by

- Chapter 3
- Section 4.1
- Chapter 5, up to Corollary 5.7(a)
- Chapter 6 up to Sect. 6.3
- Chapter 7 up to Sect. 7.2
- Chapter 8

To begin, Chap. 2 reviews background material, cites references and fixes notation.

Chapter 2
Notation and Preliminaries

The set of natural numbers $\{1, 2, \cdots , \}$ is denoted by $\mathbb{N}$, $\mathbb{N}_0 = \mathbb{N} \bigcup \{0\}$, $\mathbb{R}$ is the real numbers, $\mathbb{C}$ is the complex numbers, $\mathbb{Q}$ is the rational numbers, and extended real numbers, $\mathbb{R} \bigcup \{+\infty, -\infty\}$, are denoted by $\overline{\mathbb{R}}$. The empty set is denoted by $\emptyset$, and a set with only one element is called a singleton.

For an arbitrary set S, $\wp(S)$ denotes the collection of all its subsets, including the empty set $\emptyset$ and S itself. A set S is said to be denumerable or countable if there is an injection from S into $\mathbb{N}$, and uncountable otherwise. If S_k is denumerable for all $k \in \mathbb{N}$, then $\bigcup_k S_k$ is denumerable. If there is an injection from S into the set $\{1, 2, \cdots , K\}$ for some $K \in \mathbb{N}$, S is said to be finite, and infinite otherwise. If S is an infinite denumerable set, there is a bijection from S onto $\mathbb{N}$.

2.1 Partial Ordering and Zorn's Lemma

For a non-empty set S and $R \subset S \times S$, a relation $\preceq$ on S is defined by writing $x \preceq y$ if $(x, y) \in R$.

Definition 2.1 (*Partial and Total Ordering*) A partial ordering on S is a relation which satisfies the following axioms [12, Sect. I.2]:

(i) $x \preceq x$ for all $x \in S$,
(ii) $x \preceq y$ and $y \preceq z$ implies that $x \preceq z$ for all $x, y, z \in S$,
(iii) $x \preceq y$ and $y \preceq x$ implies $x = y$ for all $x, y \in S$.

If in addition at least one of $x \preceq y$ or $y \preceq x$ holds for every $(x, y) \in S \times S$, the partial ordering $\preceq$ is said to be a total ordering on S. If $x \preceq y$ but $x \neq y$, write $x \prec y$. $\square$

Example 2.2 The usual ordering $\leq$ on $\mathbb{R}$ is a partial ordering which is a total ordering on every subset S of $\mathbb{R}$.

© The Author(s), under exclusive license to Springer Nature Switzerland AG 2020
J. Toland, *The Dual of $L_\infty(X, \mathcal{L}, \lambda)$, Finitely Additive Measures and Weak Convergence*, SpringerBriefs in Mathematics,
https://doi.org/10.1007/978-3-030-34732-1_2

Any collection $\mathcal{S}$ of subsets of a set is partially ordered by set inclusion, i.e. $A \preceq B$ if and only if $A \subset B$, but if $\mathcal{S}$ is not a singleton it may not be totally ordered. $\square$

Definition 2.3 (*Upper Bounds and Maximal Elements*) If $\preceq$ is a partial ordering on S and $A \subset S$,

(i) $u \in S$ is an upper bound of A if $a \preceq u$ for all $a \in A$,
(ii) $m \in A$ is a maximal element of A if $m \npreceq a$ for all $a \in A$. $\square$

Lemma 2.4 (Zorn) *Suppose $\preceq$ is a partial ordering on S and every totally ordered subset $A \subset S$ has an upper bound. The S has a maximal element.*

The proof depends on an assumption that for any collection $\mathcal{T}$ of subsets of S there is a set which contains a point from each of the sets in $\mathcal{T}$. More precisely:

The Axiom of Choice
Let S be non-empty and $\mathcal{T}$ a non-empty subset of $\wp(S) \setminus \{\emptyset\}$. Then there exists a function $f : \mathcal{T} \rightarrow S$ such that $f(T) \in T$ for all $T \in \mathcal{T}$.

Conversely, the Axiom of Choice can be proved using Zorn's lemma. So these two seemingly different statements are in fact equivalent formulations of a fundamental axiom of set theory.

Definition 2.5 (*Equivalence Relation*) A relation $\sim$ on a set S is called an equivalence relation if for all $x, y, z \in S$, (i) $x \sim x$; (ii) $x \sim y \Leftrightarrow y \sim x$; (iii) $x \sim y$ and $y \sim z \Rightarrow x \sim z$. The set $[x] = \{y \in S : x \sim y\}$ is the equivalence class which contains x. $\square$

2.2 Algebras and σ-Algebras

For an arbitrary set X, a collection $\mathcal{M} \subset \wp(X)$ is an algebra if

$\emptyset$ and X are in $\mathcal{M}$,
$A \in \mathcal{M}$ if and only $X \setminus A \in \mathcal{M}$,
$\bigcup_{k=1}^{K} A_k \in \mathcal{M}$ when $A_k \in \mathcal{M}$, $k = 1, 2, \cdots, K$, $K \in \mathbb{N}$.

If $\mathcal{M}$ is an algebra and $A, B \in \mathcal{M}$, it follows that $A \cap B \in \mathcal{M}$, $A \setminus B \in \mathcal{M}$ and $\mathcal{M}$ is closed under finite unions and intersections.

A σ-algebra, denoted by $\mathcal{L}$, is an algebra with the additional property that

$A_k \in \mathcal{L}, k \in \mathbb{N}$, implies $\bigcup_{k \in \mathbb{N}} A_k \in \mathcal{L}$.

Hence, a σ-algebra is closed under complementation, and countable unions and intersections.

Examples.

For any set X, $\{\emptyset, X\}$ and $\wp(X)$ are σ-algebras.

With $X = [0, 1)$, the collection of unions of finitely many intervals of the form $[a, b), 0 \le a \le b \le 1$, is an algebra, but not a σ-algebra.

$\mathcal{M} \subset \wp(\mathbb{N})$, defined by $A \in \mathcal{M}$ if and only if either A or $\mathbb{N} \setminus A$ is finite, is an algebra. Clearly, the singletons $\{2n + 1\} \in \mathcal{M}$ for all $n \in \mathbb{N}_0$. So $\mathcal{M}$ is not a σ-algebra because $\bigcup_{n \in \mathbb{N}_0} \{2n + 1\} \notin \mathcal{M}$.

On the other hand, saying $A \in \mathcal{L}$ if and only if A or $X \setminus A$ is denumerable defines a σ-algebra $\mathcal{L}$ on any set X. (When X is uncountable, $\mathcal{L} \neq \wp(X)$.)

It is obvious that if $\mathcal{L}_\kappa$, $\kappa \in \mathcal{K}$, where $\mathcal{K}$ is an arbitrary set, are σ-algebras, $\mathcal{L} := \bigcap_{\kappa \in \mathcal{K}} \mathcal{L}_\kappa$ is also a σ-algebra. Therefore, since $\wp(X)$ is a σ-algebra, for any $\mathcal{V} \subset \wp(X)$ the intersection of all σ-algebras which contain $\mathcal{V}$ is the smallest σ-algebra containing $\mathcal{V}$, called the σ-algebra generated by $\mathcal{V}$.

In particular, an algebra $\mathcal{M}$ generates a σ-algebra, denoted by $\overline{\mathcal{M}}$ in [35]. Another special case arises when (X, ϱ) is a topological space, $\varrho \subset \wp(X)$ being the collection of open sets in X. Then the σ-algebra generated by ϱ is called the Borel σ-algebra of (X, ϱ), often denoted by $\mathcal{B}$ and elements of $\mathcal{B}$ are called Borel subsets of X.

2.3 Measurable Sets and Measurable Functions

A measurable space is a pair $(X, \mathcal{L})$ where $X \neq \emptyset$ is arbitrary and $\mathcal{L} \subset \wp(X)$ is a σ-algebra. Elements of $\mathcal{L}$ are referred to as measurable sets and a function $u : X \to \overline{\mathbb{R}}$ is said to be measurable if

$$\left\{ x : u(x) > \alpha \right\} \in \mathcal{L} \text{ for all } \alpha \in \mathbb{R}.$$

The set of measurable functions has the following properties.

(i) If u_n, $n \in \mathbb{N}$, is measurable and $u_n(x) \to u(x)$ for all $x \in X$, u is measurable.

(ii) If $\{u_n\}$ is a sequence of extended real-valued measurable functions,

$$\underline{u}(x) := \inf_n u_n(x) \text{ and } \overline{u}(x) := \sup_n u_n(x), \ x \in X, \text{ are measurable.}$$

(iii) If u is measurable, $u^\pm$ are measurable where $u^\pm(x) := \sup\{\pm u(x), 0\}$, so that $u^\pm(x) \geq 0$, $x \in X$, and

$$u = u^+ - u^- \text{ and } |u| = u^+ + u^-. \tag{2.1}$$

(iv) If u, v are real-valued and measurable, $c \in \mathbb{R}$ and $g : \mathbb{R} \to \mathbb{R}$ is continuous, then cu, $g(u) = g \circ u$, $u^\pm$, uv, $u + v$ are measurable.

(v) If u is non-negative and measurable, there is a sequence $\{\varphi_n\}$ of non-negative measurable functions on X with

(a) $0 \leq \varphi_n(x) \leq \varphi_{n+1}(x) \leq u(x)$, $x \in X$, $n \in \mathbb{N}$,
(b) $u(x) = \lim_n \varphi_n(x)$, $x \in X$,
(c) φ_n is real-valued with only finitely many values for each $n \in \mathbb{N}$.

The characteristic function χ_A of a set $A \in \mathcal{L}$, defined by

$$\chi_A(x) = \begin{cases} 1 \text{ if } x \in A \\ 0 \text{ otherwise} \end{cases},$$

is a measurable function. Finite linear combinations of characteristic functions

$$\varphi(x) = \sum_{k=1}^{K} a_k \chi_{A_k}, \quad a_k \in \mathbb{R}, \ A_k \in \mathcal{L}, \ K \in \mathbb{N}, \tag{2.2}$$

are called simple functions and are measurable. The functions φ_n in (v) above are non-negative simple functions.

2.4 Measures and Real Measures

In the measure theory literature, subtly different meanings are often assigned by different authors to the same terminology. For example, in [15, Chap. 3, p. 85] a signed measure may have infinite values, whereas in [28, Sect. 6.6, p. 119] it may not. See also Remark 2.38. For this reason, the terminology chosen for subsequent chapters is described in some detail below.

Definition 2.6 (*Measures*) In a measurable space $(X, \mathcal{L})$, a measure λ is an extended real-valued function on $\mathcal{L}$ with $\lambda(X) > 0$, $\lambda(\emptyset) = 0$, $\lambda(A) \geq 0$ for all $A \in \mathcal{L}$, and when $E_k \in \mathcal{L}$, $k \in \mathbb{N}$, and $E_i \bigcap E_j = \emptyset$, $i \neq j$,

$$\lambda\Big(\bigcup_{k \in \mathbb{N}} E_k\Big) = \sum_{k \in \mathbb{N}} \lambda(E_k). \tag{2.3}$$

When $\mathcal{L}$ is a σ-algebra this identity is referred to as the σ-additivity of λ. Some texts use the term "positive measure" for what here is referred to as a measure. □

The triple $(X, \mathcal{L}, \lambda)$ is called a measure space. A measure is said to be finite if $\lambda(X) < \infty$, and σ-finite if $X = \bigcup_{n \in \mathbb{N}} X_n$ where $X_n \in \mathcal{L}$ and $\lambda(X_n) < \infty$ for all n. In a measure space a set $E \in \mathcal{L}$ is null, written as $E \in \mathcal{N}$, if $\lambda(E) = 0$, and a property is said to hold λ-almost everywhere if the exceptional set where it does not hold is null. A measure space is said to be complete if $A \subset B \in \mathcal{N}$ implies that $A \in \mathcal{N}$.

Definition 2.7 In a measure space $(X, \mathcal{L}, \lambda)$, a sequence $\{u_k\}$ of measurable functions is said to converge in measure to a measurable function u if for all $\alpha > 0$

$$\lambda\Big(\{x : |u_k(x) - u(x)| > \alpha\}\Big) \to 0 \text{ as } k \to \infty. \quad □$$

A sequence of measurable functions which converges in measure has a subsequence which converges pointwise λ-almost everywhere [15, Thm. 2.30].

Definition 2.8 (*Atom*) $A \in \mathcal{L}$ is an atom if $A \notin \mathcal{N}$ and $A \supset B \in \mathcal{L}$ implies that either $\lambda(B) = 0$ or $\lambda(B) = \lambda(A)$. $\qquad\square$

Remark 2.9 An atom A in a σ-finite measure space has finite measure. If $\lambda(A) < \infty$ and u is a bounded measurable function let

$$a = \inf \big\{ b \in \mathbb{R} : \lambda(\{x \in A : u(x) \leq b\}) = \lambda(A) \big\}.$$

Then $\lambda(\{x \in A : u(x) \leq a\}) = \lambda(A)$ and $\lambda(\{x \in A : u(x) < a\}) = 0$, since λ is σ-additive and $\lambda(\{x \in A : u(x) \leq a - 1/k\}) = 0$ for all k. In other words $u = a$, a constant, λ-almost everywhere on A. The next result is used in Sect. 2.5. $\qquad\square$

Lemma 2.10 *If $(X, \mathcal{L}, \lambda)$ has no atoms, for all $G \in \mathcal{L}$ with $\lambda(G) > 0$ and $\epsilon > 0$ there exists a subset $E \in \mathcal{L}$ of G with $\lambda(E) \in (0, \epsilon]$.*

Proof Since there are no atoms, the given G can be replaced by a subset, also denoted by $G \in \mathcal{L}$, with $\lambda(G) \in (0, \infty)$. Let $\epsilon > 0$. Now since G is not an atom, there exists $E_1 \subset G, E_1 \in \mathcal{L}$, with $\lambda(E_1) \in (0, \lambda(G))$. If $\lambda(E_1) > \epsilon$ for all such E_1 it follows that $\lambda(E_1) > \epsilon$ and $\lambda(G \setminus E_1) > \epsilon$. Since there are no atoms, by the same argument there exists $E_2 \subset G \setminus E_1$ with $\lambda(E_2) > \epsilon$ and $\lambda(G \setminus (E_1 \bigcup E_2)) > \epsilon$. By induction there is a sequence $\{E_k\} \subset \mathcal{L}$ of disjoint subsets of G with $\lambda(E_k) > \epsilon$. Since $\lambda(G) < \infty$ this is false. Hence, there exists $E \subset G, E \in \mathcal{L}$ with $\lambda(E) \in (0, \epsilon]$. $\qquad\square$

Remark 2.11 Lemma 2.10 implies the Darboux property of atomless measures [8, Cor. 1.12.10] which is apparently stronger: If $(X, \mathcal{L}, \lambda)$ has no atoms and $\lambda(F) > 0$ for some $F \in \mathcal{L}$, then for any $a \in (0, \lambda(F))$ there exists $E \in \mathcal{L}$ with $E \subset F$ and $\lambda(E) = a$. (The prototype is due independently to Fichtenholz [13] and Sierpiński [30].) $\qquad\square$

Definition 2.12 (*Regular Borel Measure*) When $\mathcal{B}$ is a Borel σ-algebra on a topological space (X, ϱ), a measure λ on $\mathcal{B}$ is called a Borel measure. For $B \in \mathcal{B}$, a Borel measure is said to be

$$\text{outer regular on } B \text{ if } \lambda(B) = \inf \big\{ \lambda(U) : B \subset U \in \varrho \big\},$$

$$\text{inner regular on } B \text{ if } \lambda(B) = \sup \big\{ \lambda(K) : K \subset B, K \text{ compact} \big\}.$$

A Borel measure which is both outer and inner regular on every $B \in \mathcal{B}$ is called regular. $\qquad\square$

Example 2.13 (*Dirac Measures*) For any measurable space $(X, \mathcal{L})$ and $x \in X$, let δ_x be defined on $\mathcal{L}$ by $\delta_x(E) = 1$ if $x \in E$ and $\delta_x(E) = 0$ otherwise. Then δ_x is a measure, called a Dirac measure at x, and $(X, \mathcal{L}, \delta_x)$ is a finite, complete measure space in which $\{x\}$ is an atom. $\qquad\square$

Example 2.14 (*Counting Measure on* $\mathbb{N}$) Let $X = \mathbb{N}, \mathcal{L} = \wp(\mathbb{N})$ and define $\lambda(E)$ as the number of elements in $E \subset \mathbb{N}$ if E is finite, and $\lambda(E) = +\infty$ if E is infinite. Then $(\mathbb{N}, \wp(\mathbb{N}), \lambda)$ is a σ-finite, complete measure space, λ is called counting measure and every singleton $\{n\}, n \in \mathbb{N}$, is an atom. $\qquad\square$

Example 2.15 (Lebesgue Measure on $\mathbb{R}^n$) Let $\mathcal{B}$ denote the Borel subsets of $X = \mathbb{R}^n$ with the standard metric and $(\mathbb{R}^n, \mathcal{B})$ the corresponding measurable space. Then although there is a unique σ-finite measure λ' on $\mathcal{B}$ which coincides with the n-dimensional volume of balls in $\mathbb{R}^n$, it is not complete. However, a complete, σ-finite measure space $(\mathbb{R}^n, \mathcal{L}, \lambda)$ is defined as follows:

$$\mathcal{L} = \left\{ B \bigcup A' : B \in \mathcal{B},\ A' \subset B' \in \mathcal{B},\ \lambda'(B') = 0 \right\},$$
$$\lambda\left(B \bigcup A' \right) = \lambda'(B),\quad B \bigcup A' \in \mathcal{L}.$$

This is the classical Lebesgue measure space and λ is Lebesgue measure on $\mathbb{R}^n$. Thus every Borel measurable function is Lebesgue measurable, and every Lebesgue measurable function is equal almost everywhere to a Borel measurable function.

To see that $\mathbb{R}^n$ with Lebesgue measure has no atoms, suppose $E \in \mathcal{L}$ and $\lambda(E) > 0$. Now define a function $f : [0, \infty) \to \mathbb{R}$ by $f(r) = \lambda(E \cap B_r)$ where B_r denotes the ball of radius r centred at 0 in $\mathbb{R}^n$. From the σ-additivity of λ, it follows that f is continuous with $f(0) = 0$ and $f(r) \to \lambda(E) > 0$ as $r \to \infty$. So, by the intermediate value theorem, there exists $\hat{r}$ with

$$\lambda(E \cap B_{\hat{r}}) = f(\hat{r}) = \lambda(E)/2 \in (0, \lambda(E)),$$

which shows that E is not an atom.

If $\Omega \in \mathcal{L}$, a typical case being when Ω is an open subset of $\mathbb{R}^n$, the restriction of λ to the σ-algebra $\{\Omega \cap E : E \in \mathcal{L}\}$ creates a complete measure space which is denoted by $(\Omega, \mathcal{L}, \lambda)$. $\qquad\square$

Definition 2.16 *(Real Measures)* In a measurable space $(X, \mathcal{L})$, a real measure [28, Sects. 1.8 and 6.6] is a real-valued function μ (not necessarily one-signed) on $\mathcal{L}$ with $\mu(\emptyset) = 0$ and

$$\mu\left(\bigcup_{i \in \mathbb{N}} E_i \right) = \sum_{i \in \mathbb{N}} \mu(E_i) \text{ when } E_i \in \mathcal{L} \text{ and } E_i \cap E_j = \emptyset,\ i \neq j \in \mathbb{N}.$$

Since the left side is independent of the ordering of $i \in \mathbb{N}$, the sum of the series on the right is the same if i is replaced by $\sigma(i)$, where σ is a permutation of $\mathbb{N}$. Hence, the series is absolutely convergent. $\qquad\square$

For a real measure μ and $E \in \mathcal{L}$ let

$$|\mu|(E) = \sup\left\{ \sum_{i=1}^{\infty} |\mu(E_i)| : E = \bigcup_{i=1}^{\infty} E_i,\ E_i \in \mathcal{L},\ E_j \cap E_j = \emptyset,\ i \neq j \in \mathbb{N} \right\}.$$

Then $|\mu(E)| \leq |\mu|(E)$, $E \in \mathcal{L}$, and $|\mu|$ is a measure in the sense of Definition 2.6 [28, Thm. 6.2] with $|\mu|(X) < \infty$ [28, Thm. 6.4]. Hence $\mu^{\pm}$ are real measures where

$$0 \leq \mu^+ := \tfrac{1}{2}(|\mu| + \mu) \text{ and } 0 \leq \mu^- := \tfrac{1}{2}(|\mu| - \mu), \tag{2.4}$$

whence

$$\mu = \mu^+ - \mu^- \text{ and } |\mu| = \mu^+ + \mu^-, \ 0 \leq \mu^\pm(E) \leq |\mu|(X) < \infty, \ E \in \mathcal{L}.$$

$|\mu|$ is called the total variation of μ, $\mu^\pm$ are the positive and negative parts of μ, and $\mu = \mu^+ - \mu^-$ is the Jordan decomposition of μ.

Theorem 2.17 (Hahn Decomposition [28, Thm 6.14]) *Let μ be a real measure on a σ-algebra $\mathcal{L}$. Then there exist $A^\pm \in \mathcal{L}$ such that $A^+ \bigcup A^- = X$, $A^+ \bigcap A^- = \emptyset$ and for $E \in \mathcal{L}$,*

$$\mu^+(E) = \mu(A^+ \bigcap E), \ \ \mu^-(E) = -\mu(A^- \bigcap E)$$

where $\mu^\pm$ are defined in (2.4).

Definition 2.18 (*Absolute Continuity and Singularity*) A real measure μ is absolutely continuous with respect to a measure λ, written as

$$\mu \ll \lambda \text{ if and only if } \lambda(E) = 0 \text{ implies } |\mu|(E) = 0$$

and singular with respect to λ, written as

$$\lambda \perp \mu \text{ if and only if } \lambda(E) + |\mu|(X \setminus E) = 0 \text{ for some } E \in \mathcal{L}. \qquad \square$$

This notation will be generalised in Definition 4.6 and Remark 4.7 to accommodate finitely additive measures that are not σ-additive. In Chap. 4, real measures on $\mathcal{L}$ are seen as examples of finitely additive measures, in which context they are denoted by $\Sigma(\mathcal{L})$.

Remark 2.19 Note that in [15, Chap. 7] the term "signed measure" allows μ to take one, but not both, of the values $\pm\infty$. Thus in that terminology a real measure would be a signed measure but a signed measure might not be a real measure. $\qquad \square$

Example 2.20 Any finite linear combination of finite measures is a real measure. A class of real measures that are absolutely continuous with respect to λ will be defined in Remark 2.27. $\qquad \square$

Definition 2.21 (*Regular Real Borel Measures*) Let $\mathcal{B}$ denote the Borel σ-algebra of a locally compact Hausdorff space (X, ϱ). Then a real Borel measure μ is said to be regular if both the measures $\mu^\pm$ are regular in the sense of Definition 2.12: for all $B \in \mathcal{B}$

$$\mu^\pm(B) = \inf \left\{ \mu^\pm(U) : B \subset U \in \varrho \right\} = \sup \left\{ \mu^\pm(K) : K \subset B, K \text{ compact} \right\},$$

where $\mu^\pm$ are defined by (2.4). $\qquad \square$

2.5   Splitting Families of Measurable Sets

This section concerns measurable sets which in an arbitrary measure space $(X, \mathcal{L}, \lambda)$ split families of measurable sets in the following sense.

Definition 2.22 (*Splitting Measurable Sets*) $A \in \mathcal{L}$ splits $\mathcal{G} \subset \mathcal{L} \setminus \mathcal{N}$ if $\lambda(G \cap A) > 0$ and $\lambda(G \setminus A) > 0$ for all $G \in \mathcal{G}$. $\square$

Remark 2.23 If A splits $\mathcal{G}$, then $X \setminus A$ also splits $\mathcal{G}$. Moreover, if $\mathcal{G}_1, \mathcal{G}_2 \subset \mathcal{L} \setminus \mathcal{N}$ are such that for each $G_1 \in \mathcal{G}_1$ there exists $G_2 \in \mathcal{G}_2$ with $G_2 \subset G_1$, then A splits $\mathcal{G}_1$ if A splits $\mathcal{G}_2$. $\square$

The main result, Theorem 2.25, depends on Baire's category theorem. To set the scene define an equivalence relation $\sim$ on $\mathcal{L}$ by writing $E \sim F$ if an only if $E \Delta F$ is null, i.e. the symmetric difference $E \Delta F = (E \setminus F) \bigcup (F \setminus E) \in \mathcal{N}$. For $E, F \in \mathcal{L}$ let

$$\partial(E, F) = \tan^{-1}\left(\lambda(E \Delta F)\right) = \tan^{-1}\left(\int_X |\chi_E - \chi_F| \, d\lambda\right), \qquad (2.5)$$

where χ_A is the characteristic function of $A \in \mathcal{L}$ and $\tan^{-1}$ maps $\overline{\mathbb{R}}$ onto $[-\pi/2, \pi/2]$. Since $L_1(X, \mathcal{L}, \lambda)$ is a Banach space and since a convergent sequence in $L_1(X, \mathcal{L}, \lambda)$ has a subsequence that converges pointwise λ-almost everywhere, it is immediate that the space $(\mathcal{L}, \partial)$ is a complete metric space. Let $\hat{\mathcal{L}} = \left\{ G \in \mathcal{L} : \lambda(G) \in (0, \infty) \right\}$ and for $G \in \hat{\mathcal{L}}$ let

$$\mathcal{F}(G) = \left\{ E \in \mathcal{L} : \lambda(E \bigcap G) \in \{0, \lambda(G)\} \right\}.$$

The following simple adaption of Kirk's argument [22] yields Theorem 2.25, Lemma 7.9 and Example 9.11.

Lemma 2.24 *For any* $G \in \hat{\mathcal{L}}$ *the set* $\mathcal{F}(G)$ *is closed with empty interior in* $(\mathcal{L}, \partial)$. *In other words* $\mathcal{F}(G)$ *is nowhere dense in* $(\mathcal{L}, \partial)$.

Proof For $G \in \hat{\mathcal{L}}$ let $g : \mathcal{L} \to \mathbb{R}$ be given by $g(E) = \lambda(G \bigcap E)$. Then g is continuous because for $E_1, E_2 \in \mathcal{L}$,

$$|g(E_1) - g(E_2)| = \left| \int_X \left(\chi_{G \cap E_1} - \chi_{G \cap E_2}\right) d\lambda \right|$$

$$\leq \int_X |\chi_{E_1} - \chi_{E_2}| \, d\lambda = \tan\left(\partial(E_1, E_2)\right).$$

So $\mathcal{F}(G) = g^{-1}\{0, \lambda(G)\}$ is closed in $(\mathcal{L}, \partial)$. To see that $\mathcal{F}(G)$ is nowhere dense let $E \in \mathcal{F}(G)$ and $\epsilon > 0$ be arbitrary. By Lemma 2.10 there exists $E_\epsilon \subset G$ with $E_\epsilon \in \mathcal{L}$ and $0 < \lambda(E_\epsilon) < \min\{\tan \epsilon, \lambda(G)\}$.

     If $\lambda(E \bigcap G) = 0$ let $E^+ = E \bigcup E_\epsilon$, so that $0 < \lambda(E^+ \bigcap G) = \lambda(E_\epsilon) < \lambda(G)$ and $\partial(E, E^+) = \tan^{-1}\left(\lambda(E_\epsilon \setminus E)\right) \leq \tan^{-1}\left(\lambda(E_\epsilon)\right) < \epsilon$.

In the other case, when $\lambda(E \cap G) = \lambda(G)$, let $E^- = E \setminus E_\epsilon$, so that

$$\lambda(E^- \cap G) = \lambda((E \cap G) \setminus E_\epsilon)) = \lambda(G \setminus E_\epsilon) = \lambda(G) - \lambda(E_\epsilon) \in (0, \lambda(G))$$

and $\partial(E, E^-) = \tan^{-1}\left(\lambda(E \setminus E^-)\right) = \tan^{-1}\left(\lambda(E \cap E_\epsilon)\right) < \epsilon.$

In both cases $E^\pm \in \mathcal{L}$ with $\partial(E^\pm, E) < \epsilon$ but $E^\pm \notin \mathcal{F}(G)$. Hence, $\mathcal{F}(G)$ is closed with empty interior when $G \in \hat{\mathcal{L}}$, as required. $\qquad\square$

Since Lebesgue measure on $\mathbb{R}^n$ has no atoms and counting measure on $\mathbb{N}$ has infinitely many atoms, the following result applies to the former but not the latter.

Theorem 2.25 *If $(X, \mathcal{L}, \lambda)$ has no atoms and $\mathcal{G}$ is a denumerable subset of $\mathcal{L} \setminus \mathcal{N}$, there exists an infinite sequence $\{A_i\}$ in $\mathcal{L}$ of mutually disjoint sets such that A_i splits $\mathcal{G}$ for all $i \in \mathbb{N}$.*

Proof Let $\mathcal{G} = \{G_j : j \in \mathbb{N}\}$. Since there are no atoms, each G_j has a subset $\tilde{G}_j$ with $\lambda(\tilde{G}_j) \in (0, \infty)$. So, by Remark 2.23, it suffices to assume that $G_j \in \hat{\mathcal{L}}$ for all j. Then, by Baire's category theorem, $\mathcal{L} \neq \bigcup_{j \in \mathbb{N}} \mathcal{F}(G_j) \subset \mathcal{L}$ since, by Lemma 2.24, each $\mathcal{F}(G_j)$ is nowhere dense in the complete metric space $(\mathcal{L}, \partial)$. If $A_1 \in \mathcal{L} \setminus \bigcup_{j \in \mathbb{N}} \mathcal{F}(G_j)$ then, from the definition of $\mathcal{F}(G_j)$, A_1 splits $\mathcal{G}$.

Now let $X_1 = X \setminus A_1, \mathcal{L}_1 = \{E \setminus A_1 : E \in \mathcal{L}\}$ and $\mathcal{G}_1 = \{G \setminus A_1 : G \in \mathcal{G}\}$. Then $\mathcal{G}_1 \subset \mathcal{L}_1 \setminus \mathcal{N}$ satisfy the hypotheses of the theorem in the measure space $(X_1, \mathcal{L}_1, \lambda)$. So there exists $A_2 \subset X \setminus A_1, A_2 \in \mathcal{L}_1$, which splits $\mathcal{G}_1$. By construction $A_2 \cap A_1 = \emptyset$ and A_2 splits $\mathcal{G}$ by Remark 2.23. Iteration of this argument yields $\{A_i : i \in \mathbb{N}\} \subset \mathcal{L}$ as in the statement of the theorem. $\qquad\square$

2.6 Integration

The Lebesgue integral (which may be infinite) of a non-negative simple function (2.2) with respect to a measure γ on $\mathcal{L}$ is defined in an obvious way by

$$\int_X \varphi \, d\gamma = \sum_{k=1}^K a_k \gamma(A_k),$$

where $a_k \geq 0$ and $\gamma(A) \geq 0$ for all $A \in \mathcal{L}$. Since the functions φ_n in Sect. 2.3 (v) are non-negative simple functions, the integral of a non-negative measurable function u with respect to non-negative γ may be defined by

$$\int_X u \, d\gamma = \lim_{n \to \infty} \int_X \varphi_n \, d\gamma. \tag{2.6}$$

Because γ is σ-additive it is easily seen that the limit on the right is independent of the sequence $\{\varphi_n\}$ which satisfies (v)(a)–(c).

Remark 2.26 Note that for $u \in L_\infty(X, \mathcal{L}, \lambda)$ this definition coincides with the definition in Sect. 6.1 when γ is σ-additive, but not when γ is only finitely additive. This is because

$$\chi_{(0, 1-1/k)}(x) \nearrow 1 \text{ for all } x \in (0, 1),$$

but (†) in the Preface (established in Remark 6.1) contradicts (2.6). Indeed (†) shows that the limit in (2.6) may not be independent of the sequence $\{\varphi_n\}$ when γ is not σ-additive.   □

Finally, if γ is either a measure (in which case $\gamma^- = 0$) or an arbitrary real measure, a measurable function u is said to be Lebesgue integrable with respect to γ if and only if

$$\int_X u^+ \, d\gamma^+, \quad \int_X u^- \, d\gamma^+, \quad \int_X u^+ \, d\gamma^-, \quad \int_X u^- \, d\gamma^- \text{ are all finite,}$$

in which case

$$\int_X u \, d\gamma := \int_X u^+ \, d\gamma^+ - \int_X u^- \, d\gamma^+ - \int_X u^+ \, d\gamma^- + \int_X u^- \, d\gamma^-.$$

It follows that u is integrable with respect to γ if and only if $|u|$ is integrable with respect to $|\gamma|$ and

$$\left| \int_X u \, d\gamma \right| \leq \int_X |u| \, d|\gamma| < \infty.$$

Moreover

$$\int_X |u| \, d\lambda = 0 \text{ if and only if } u \text{ is zero } \lambda\text{-almost everywhere.} \tag{2.7}$$

Remark 2.27 When $(X, \mathcal{L}, \lambda)$ is a measure space and $h : X \to \mathbb{R}$ is any integrable function, a real measure μ_h is defined on $\mathcal{L}$ by

$$\mu_h(E) = \int_X h\chi_E \, d\lambda, \quad E \in \mathcal{L}, \text{ and } \mu_h \ll \lambda.$$

In fact part (b) of the next result indicates that all real measures μ that are absolutely continuous with respect to λ arise in this way.   □

Theorem 2.28 (Lebesgue–Radon–Nikodym [15, Thm. 3.8], [28, Sect. 6.10]) *For a σ-finite measure λ and a real measure μ on $\mathcal{L}$ there is*
   (a) a unique pair μ_a, μ_s of real measures with $\mu = \mu_a + \mu_s$, $\mu_a \ll \lambda$ and $\mu_s \perp \lambda$ (Definition 2.18), and if μ is positive then so are μ_a and μ_s, and

(b) a unique function h which is integrable with respect to λ with the property that $\mu_a(E) = \int_E h \, d\lambda$ for all $E \in \mathcal{L}$.

Part (a) is commonly referred to as the Lebesgue decomposition of μ relative to λ and (b) is the Radon–Nikodym theorem.

2.7 Function Spaces

For spaces of pth-power integrable functions, the notation is that of [6, 12, 35], namely, L_p rather than L^p, because for dual spaces L_p^* is easier to manage than L^{p*}. With the exception of Sect. 9.4 only spaces of real-valued functions will be considered.

L_p**-Spaces.** When $(X, \mathcal{L}, \lambda)$ is a measure space an equivalence relation $\sim$ is defined on measurable functions by $u \sim v$ if and only if $u = v$ λ-almost everywhere. For convenience with notation, it is usual to denote the equivalence class $[u]$ (Definition 2.5) simply as u and to refer to it as a function rather than an equivalence class of functions.

Let $L_p(X, \mathcal{L}, \lambda)$, $p \in [1, \infty)$, denote the (equivalence classes of) measurable functions u with $\int |u|^p \, d\lambda < \infty$. Then it is well known that with norm

$$\|u\|_p := \left(\int |u|^p \, d\lambda \right)^{1/p}, \quad p \in [1, \infty),$$

$L_p(X, \mathcal{L}, \lambda)$ is a real Banach space. It is obvious that $\|u_k - u\|_p \to 0$ as $k \to \infty$ for any $p \in [1, \infty)$ implies that $u_k \to u$ in measure (Definition 2.7), and hence for a subsequence that $u_{k_j}(x) \to u(x)$ λ-almost everywhere.

Example 2.29 Let $X = (0, 1)$ with the usual metric and Lebesgue measure and consider the intervals

$$I_1 = \left(0, 1\right);$$

$$I_2 = \left(0, \frac{1}{2}\right); \ I_3 = \left(\frac{1}{2}, 1\right);$$

$$I_4 = \left(0, \frac{1}{3}\right); \ I_5 = \left(\frac{1}{3}, \frac{2}{3}\right); \ I_6 = \left(\frac{2}{3}, 1\right);$$

$$I_7 = \left(0, \frac{1}{4}\right); \ I_8 = \left(\frac{1}{4}, \frac{1}{2}\right); \ I_9 = \left(\frac{1}{2}, \frac{3}{4}\right); \ I_{10} = \left(\frac{3}{4}, 1\right);$$

and so on. Note that $\lambda(I_k) \leq 1/(n + 1)$ if $k > n(n + 1)/2$ (the sum of the first n natural numbers). Let $u_k = \chi_{I_k}$. Then $u_k \to 0$ in $L_p(0, 1)$ for all $p \in [1, \infty)$, from which it follows that $u_k \to 0$ in measure, and hence there exists a subsequence with $u_{k_j}(x) \to 0$ for λ-almost all $x \in (0, 1)$. However, it is obvious from the construction that

$$\limsup_{k\to\infty} u_k(x) = 1 \text{ for } \lambda\text{-almost all } x \in (0,1). \quad \square$$

A measurable function u is said to be essentially bounded if $|u(x)| \le c$, a constant, λ-almost everywhere. With norm

$$\|u\|_\infty = \inf\left\{c > 0 : |u(x)| \le c \; \lambda\text{-almost everywhere}\right\} \tag{2.8a}$$

the space $L_\infty(X, \mathcal{L}, \lambda)$ of (equivalence classes of) essentially bounded measurable functions is a real Banach space. Let

$$[u]_\infty^+ = \inf\left\{c : u(x) \le c\lambda\text{-almost everywhere}\right\}, \tag{2.8b}$$

$$[u]_\infty^- = \sup\left\{c : u(x) \ge c\lambda\text{-almost everywhere}\right\}. \tag{2.8c}$$

Then $[u]^- \le u(x) \le [u]^+$ for λ-almost all $x \in X$.

Definition 2.30 (*Local Compactness* [28, Defn. 2.3]) A topological space (X, ϱ) is locally compact if for every $x \in X$ there is an open set G with $x \in G$ and the closure $\overline{G}$ of G is compact. $\qquad\square$

Definition 2.31 (σ-*Compactness* [28, Defn. 2.16]) A topological space (X, ϱ) is σ-compact if $X = \bigcup_{k\in\mathbb{N}} K_k$ for a denumerable family of compact sets K_k. $\qquad\square$

Definition 2.32 (*One-point Compactification* [21, p. 150]) The one-point compactification of a topological space (X, ϱ) is the compact space $(X_\infty, \varrho_\infty)$ defined by putting $X_\infty = X \bigcup \{x_\infty\}$ where $x_\infty \notin X$ (x_∞ is referred to as the point at infinity) and saying $G \subset X_\infty$ is open, i.e. $G \in \varrho_\infty$, if either $G \subset X$ and $G \in \varrho$, or $x_\infty \in G$ and $G = \{x_\infty\} \bigcup (X \setminus K)$ for a compact set K in (X, ϱ). $\qquad\square$

Remark 2.33 $(X_\infty, \varrho_\infty)$ is Hausdorff if and only if (X, ϱ) is locally compact and Hausdorff, and (X, ϱ) is compact if and only if $\{x_\infty\}$ is isolated (both open and closed) in $(X_\infty, \varrho_\infty)$ [21, loc. cit.]. $\qquad\square$

Continuous Function Spaces

For a locally compact Hausdorff space (X, ϱ), let $C_0(X, \varrho)$ denote the linear space of real-valued continuous functions v with the property that for all $\epsilon > 0$ there exists a compact set $K \subset X$ such that $|v(x)| < \epsilon$ for all $x \in X \setminus K$. With the maximum norm

$$\|v\|_\infty = \max_{x\in X} |v(x)|, \quad v \in C_0(X, \varrho), \tag{2.9}$$

$C_0(X, \varrho)$ is a real Banach space. When X is compact $C_0(X, \varrho)$ consists of all the real-valued continuous functions on X. When (X, ϱ) is not compact $C_0(X, \varrho)$ consists of the restrictions to X of real-valued functions that are continuous on the compact space $(X_\infty, \varrho_\infty)$ and zero at x_∞.

Remark 2.34 In a Hausdorff topological space (X, ϱ) with Borel σ-algebra $\mathcal{B}$ every singleton $\{a\}$ is closed, and hence $\chi_{\{a\}}$, $a \in X$, is Borel measurable. Now suppose that all Borel measurable functions are continuous. Then $\{a\} = \chi_{\{a\}}^{-1}((0, 2))$ is open, because $\chi_{\{a\}}$ is continuous. Therefore, all subsets of X are open, and hence $\varrho = \wp(X)$ (ϱ is the discrete topology) when all measurable functions are continuous. Conversely, if ϱ is the discrete topology every function is continuous. An important case is given by Example 2.14. $\qquad\square$

The following observation is a corollary of Lusin's theorem.

Theorem 2.35 ([28, Cor. 2.24]) *Suppose (X, ϱ) is a locally compact Hausdorff space, $\lambda(X) < \infty$ and λ is a complete regular real Borel measure. Then for $u \in L_\infty(X, \mathcal{B}, \lambda)$ there exists a sequence $\{g_k\}$ of continuous functions on (X, ϱ) such that each g_k has compact support (the closure of $\{x : g_k(x) \neq 0\}$ is compact), $\|g_k\|_\infty \leq \|u\|_\infty$ and $u(x) = \lim_{k\to\infty} g_k(x)$ for λ-almost all $x \in X$.*

2.8 Dual Spaces and Measures

Given a normed linear space $(V, \|\cdot\|)$, a metric is defined on V by $\rho(x, y) = \|x - y\|$, $x, y \in V$, and V is a Banach space if (V, ρ) is complete (i.e. Cauchy sequences converge). A linear function $f : V \to \mathbb{R}$ is continuous if and only if there is a constant M such that $|f(x)| \leq M\|x\|$ for all $x \in V$. The set of all such continuous linear functions, which are usually called bounded linear functionals, forms a linear space V^* over $\mathbb{R}$, and V^*, the dual space of V, is a Banach space (whether or not V is a Banach space) when endowed with the norm

$$\|f\|_{V^*} = \sup\left\{|f(x)| : \|x\| \leq 1, \ x \in V\right\}.$$

In the special case when V is a space of real-valued functions v on X there has historically been a great deal of interest in describing the action of $f \in V^*$ on $v \in V$ as an integral of v over X. Indeed much of the subsequent chapters concern the consequences of such a description of $L_\infty(X, \mathcal{L}, \lambda)^*$. The classical result for the dual space $L_p(X, \mathcal{L}, \lambda)^*$, $p \in [1, \infty)$, a Banach space with norm

$$\|f\|_{L_p(X,\mathcal{L},\lambda)^*} = \sup\left\{|f(u)| : \|u\|_p \leq 1, \ u \in L_p(X, \mathcal{L}, \lambda)\right\}, \qquad (2.10)$$

is the following.

Theorem 2.36 ([15, Thm. 6.15]) *For $1 \leq p < \infty$ let $f \in L_p(X, \mathcal{L}, \lambda)^*$. Then for $p > 1$ there exists a unique $g \in L_q(X, \mathcal{L}, \lambda)$ such that for all $u \in L_p(X, \mathcal{L}, \lambda)$*

$$f(u) = \int_X ug\,d\lambda \text{ where } q = \frac{p}{p-1} \text{ and } \|f\|_{L_p(X,\mathcal{L},\lambda)^*} = \|g\|_q.$$

If $(X, \mathcal{L}, \lambda)$ is σ-finite, for $p = 1$ there exists $g \in L_\infty(X, \mathcal{L}, \lambda)$ such that

$$f(u) = \int_X u\, g\, d\lambda \text{ for all } u \in L_1(X, \mathcal{L}, \lambda) \text{ and } \|f\|_{L_\infty(X,\mathcal{L},\lambda)^*} = \|g\|_\infty.$$

Therefore, $L_p(X, \mathcal{L}, \lambda)^*$ and $L_q(X, \mathcal{L}, \lambda)$ are isometrically isomorphic when $p \in (1, \infty)$ and $p^{-1} + q^{-1} = 1$, and also in the limiting case $p = 1$ and $q = \infty$ if $(X, \mathcal{L}, \lambda)$ is σ-finite. For a necessary and sufficient for $L_\infty(X, \mathcal{L}, \lambda)$ to be the dual space of $L_1(X, \mathcal{L}, \lambda)$, see [16, Cor. 2.41].

The dual space $C_0(X, \varrho)^*$ is a Banach space with norm

$$\|f\|_{C_0(X,\varrho)^*} = \sup\Big\{|f(v)| : \|v\|_\infty \leq 1, \ v \in C_0(X, \varrho)\Big\}. \tag{2.11}$$

Theorem 2.37 (Riesz [28, Thm. 6.19]) *For $f \in C_0(X, \varrho)^*$ there exists a unique, regular, real Borel measure μ on X (Definition 2.21) with*

$$f(v) = \int_X v\, d\mu \text{ for all } v \in C_0(X, \varrho). \tag{2.12}$$

Moreover, $|\mu|(X) = \|f\|_{C_0(X,\varrho)^*} < \infty$ and $f(v) \geq 0$ when $0 \leq v \in C_0(X, \varrho)$ implies that $\mu \geq 0$ on Borel sets.

The space of all regular, real Borel measures on $\mathcal{B}$ is denoted by $C_0^*(X, \tau)$ and for every $\mu \in C_0^*(X, \varrho)$, (2.12) defines $f \in C_0(X, \varrho)^*$.

Remark 2.38 Although the statements are different because the nomenclature is different, the Riesz representation theorems in [15, Thm. 7.17] and [28, Thm. 6.19] are equivalent to Theorem 2.37. □

2.9 Functional Analysis

Theorem 2.39 (Hahn–Banach [12, p. 62, II.3.10]) *On a real linear space V let $p : V \to \mathbb{R}$ be sublinear:*

$$p(x + y) \leq p(x) + p(y), \quad p(\alpha x) = \alpha p(x) \text{ for all } x, y \in V, \ \alpha \in [0, \infty),$$

and suppose $f : Y \to \mathbb{R}$ is linear with $f(y) \leq p(y)$ for all $y \in Y$, where Y is a linear subspace of V. Then there is a linear function $\tilde{f} : V \to \mathbb{R}$ with

$$\tilde{f}(y) = f(y) \text{ for all } y \in Y \text{ and } \tilde{f}(x) \leq p(x) \text{ for all } x \in V.$$

It follows that $-p(-x) \leq \tilde{f}(x) \leq p(x)$ for all $x \in V$.

$\tilde{f}$ is known as as a Hahn–Banach extension of f. The following is immediate from the fact that $p(x) = \|x\|$ is sublinear on a normed linear space.

Corollary 2.40 *Let Y be a closed linear subspace of a normed linear space V and let $v_0 \in V \setminus Y$. Then there exists $f \in V^*$ such that $f(v_0) = 1$ and $f(y) = 0$ for all $y \in Y$.*

Proof Let $Z = \{y + \alpha v_0 : y \in Y, \alpha \in \mathbb{R}\}$, a linear subspace of V. Since $v_0 \neq 0$, $z \in Z$ determines a unique $\alpha \in \mathbb{R}$ and $y \in Y$ such that $z = y + \alpha v_0$ and since Y is closed, a bounded linear functional g is defined on Z by

$$g(z) = g(y + \alpha v_0) = \alpha, \quad z \in Z.$$

By Theorem 2.39 (Hahn–Banach), there exists $f \in V^*$ which coincides with g on Z. Since $g(v_0) = 1$, f has the required properties. $\qquad\square$

Definition 2.41 (*Banach Limits* [5, p. 34, Chap. II]) As in Example 2.14 and Remark 2.34, let $\ell_\infty(\mathbb{N})$ denote the space of bounded, real sequences indexed by $\mathbb{N}$. Now let $c(\mathbb{N})$ be the subspace of convergent sequences and let $l(u)$ denote the limit of $\{u(k)\} \in c(\mathbb{N})$. Since $l : c(\mathbb{N}) \to \mathbb{R}$ is positive, bounded and linear, there exists [11, Thm. III.7.1] a positive bounded linear functional L on $\ell_\infty(\mathbb{N})$ such that for $u \in \ell_\infty(\mathbb{N})$,

$$L(u) = l(u), \; u \in c(\mathbb{N}), \quad L(u) = L(u^n) \text{ where } u^n(k) = u(k+n), \; n \in \mathbb{N},$$

and $\|L\|_{\ell_\infty(\mathbb{N})} = \|l\|_{c(\mathbb{N})} = 1$. $L : \ell_\infty(\mathbb{N}) \to \mathbb{R}$ is called a Banach limit; see [31] for development of the theory. For an account based on finitely additive measures, see Remark 6.5. $\qquad\square$

Let $C(\mathcal{Z})$ with the maximum norm denote the Banach space of real-valued continuous functions on a compact Hausdorff space $(\mathcal{Z}, \varrho)$.

Theorem 2.42 (Stone–Weierstrass [11, Thm. V.8.1], [15, Thm. 4.45]) *Suppose $\mathcal{A} \subset C(\mathcal{Z})$ is a linear space in which (i) $f, g \in \mathcal{A}$ implies the product $fg \in \mathcal{A}$; (ii) the constant function $1 \in \mathcal{A}$; and (iii) for $z_1 \neq z_1 \in \mathcal{Z}$ there exists $f \in \mathcal{A}$ such that $f(z_1) \neq f(z_2)$. Then $\overline{\mathcal{A}} = C(\mathcal{Z})$, where $\overline{\mathcal{A}}$ is the closure of $\mathcal{A}$ in $C(\mathcal{Z})$.*

Theorem 2.43 (Weierstrass) *For a continuous real-valued function f on $[a, b]$ and $\epsilon > 0$ there exists a polynomial p on $[a, b]$ with $|f(x) - p(x)| < \epsilon$ for all $x \in [a, b]$.*

Definition 2.44 (*Weak Convergence*) In a normed linear space V a sequence $\{y_k\}$ converges weakly to y_0 in V, written as $y_k \rightharpoonup y_0$, if $v^*(y_k) \to v^*(y_0)$ for all $v^* \in V^*$. $\qquad\square$

By the uniform boundedness principle (also known as the Banach–Steinhaus theorem) [28, Thm. 5.8], $y_k \rightharpoonup y_0$ in V implies $\{\|y_k\|\}$ is bounded.

Theorem 2.45 (Mazur [23, Chap. 10, Thm. 6]) *If $\{y_k\} \subset K$, where K is closed and convex in a normed linear space, and $y_k \rightharpoonup y_0$, then $y_0 \in K$.*

Corollary 2.46 *If $y_k \rightharpoonup y_0$ in a normed linear space V and $\{k_j\} \subset \mathbb{N}$ is strictly increasing, there exists $\{\overline{y}_i\}$ with $\|\overline{y}_i - y_0\| \to 0$ as $i \to \infty$, where*

$$\overline{y}_i = \sum_{j=1}^{m_i} \gamma_j^i y_{k_j}, \quad \gamma_j^i \in [0, 1] \text{ and } \sum_{j=1}^{m_i} \gamma_j^i = 1, \text{ for some } m_i \in \mathbb{N}.$$

Proof For the sequence $\{k_j\} \subset \mathbb{N}$ let K be the closure in V of the set of convex combinations of elements of $\{y_{k_j} : j \in \mathbb{N}\}$. Since $y_{k_j} \rightharpoonup y_0$ and $y_{k_j} \in K$, which is closed and convex, $y_0 \in K$ follows from Theorem 2.45. $\square$

Definition 2.47 (*Weak* Topology on V^* [11, Chap. IV, Example 1.8]) For a normed linear space V a set $U \subset V^*$ is open in the weak* topology if and only if for every $u^* \in U$ there exists $v_1, \cdots, v_n \in V$ and $\epsilon > 0$ such that

$$\bigcap_{j=1}^{n} \left\{ v^* \in V^* : |(v^* - u^*)(v_j)| < \epsilon \right\} \subset U.$$ $\square$

The collection

$$\mathcal{V}_{\epsilon, v, u^*} = \left\{ v^* \in V^* : |(v^* - u^*)(v)| < \epsilon \right\}, \quad \epsilon > 0, \ v \in V, \ u^* \in V^*$$

is said to form a sub-base for the weak* topology on V^*.

Theorem 2.48 (*Alaoglu* [11, Thm. V.3.1]) *For a normed linear space V, $B^* = \{v^* \in V^* : \|v^*\|_{V^*} \leq 1\}$, the closed unit ball in V^*, is weak* compact.*

Definition 2.49 (*Weak* Convergence in V^*) For a normed linear space V a sequence $\{v_k^*\} \subset V^*$ converges weak* to $v_0^* \in V^*$, written as $v_k^* \overset{*}{\rightharpoonup} v_0^*$, if $v_k^*(v) \to v_0^*(v)$ for all $v \in V$. $\square$

Note that when V is a Banach space, weak* convergent sequences in V^* are bounded, but not necessarily if V is not a Banach space.

Definition 2.50 (*Weak* Sequential Compactness in V^*) When V is a normed linear space $K \subset V^*$ is weak* sequentially compact if every sequence $\{v_k^*\} \subset K$ has a subsequence which converges weak* to a point $v_0^* \in K$. $\square$

Remark 2.51 If V is separable (has a denumerable dense subset) the closed unit ball in V^* with the weak* topology is metrizable [11, Thm. V.5.1], and hence [23, Chap. 10, Thm. 12] weak* sequentially compact. $\square$

Definition 2.52 (*Extreme Points*) If C is a convex subset of a linear space, $c \in C$ is an extreme point of C if $c = \alpha c_1 + (1 - \alpha)c_2$ for $c_1, c_2 \in C$ and $\alpha \in (0, 1)$ implies that $c = c_1 = c_2$. $\square$

For example, the four vertices of a closed square in the plane are its extreme points, and there are no others. Every point of the unit circle is an extreme point of the closed unit disc in the plane.

Theorem 2.53 (Krein–Milman [11, Thm. V.7.4]) *If K is non-empty, compact and convex in a locally convex space, K is the closed convex hull of its extreme points. In other words, K is the closure of the intersection of all convex sets which contain the extreme points of K. (Equivalently, K is the intersection of all closed convex sets which contain the extreme points of K.)*

The next result is a consequence of Theorem 2.48 (Alaoglu) and Theorem 2.53 (Krein–Milman), since V^* with the weak* topology is a locally convex topological space when V is a normed linear space.

Corollary 2.54 *When V is a normed linear space the closed unit ball B^* in V^* is the weak* closed convex hull of its extreme points (i.e. B^* is the intersection of all weak* closed convex set which contains the extreme points of B^*). In particular, the set of extreme points of B^* is non-empty.*

The final theorem of this section says that when V is a Banach space the extreme points of B^* satisfy property (W) in the Introduction.

Theorem 2.55 (Rainwater[1] [26]) *In a Banach space, $x_k \rightharpoonup x$ if and only if $\{x_k\}$ is bounded and $x^*(x_k) \to x^*(x)$ for all extreme points x^* in the closed unit ball of the dual space.*

2.10 Point Set Topology

Definition 2.56 (*Base and Sub-base for a Topology*) In a topological space (X, ϱ) a collection $\mathcal{G}$ of open sets such that every non-empty open set is a union of sets in $\mathcal{G}$ is called a base for the topology ϱ. A sub-base for ϱ is a collection $\mathcal{G}$ of open sets with the property that the finite intersections of elements of $\mathcal{G}$ form a base for ϱ. □

Remark 2.57 Note that for any collection $\mathcal{T} \subset \wp(X)$ with the property that $X = \bigcup_{G \in \mathcal{T}} G$, there is a unique topology τ for which $\mathcal{T}$ is a sub-base. □

Definition 2.58 (*Separable*) A topological space is separable if it has a denumerable dense subset. □

Definition 2.59 (*First Axiom of Countability*) A topological space (X, ϱ) satisfies the first axiom of countability if for every $x \in X$ there is a denumerable collection $\mathcal{G}_x$ of open sets G with $x \in G$ such that $x \in U \in \varrho$ implies that $x \in G \subset U$ for some $G \in \mathcal{G}_x$. The collection $\mathcal{G}_x$ is called a local base at x for the topology ϱ. □

---

[1]Many internationally renowned mathematicians have published under the name John Rainwater, see https://en.wikipedia.org/wiki/John_Rainwater.

Definition 2.60 (*Complete Separability or Second Axiom of Countability*) A space is said to be completely separable (or to satisfy the second axiom of countability) if its topology has a denumerable base.   □

A completely separable space satisfies the first axiom of countability and is separable. In metric spaces separability and completely separability are equivalent.

If, for any pair of disjoint closed sets F_i, $i = 1, 2$, in a topological space there exists a pair of disjoint open sets G_i with $F_i \subset G_i$, the space is normal; if this holds for any closed F_1 when F_2 is a singleton the space is regular, and if it holds when both F_i are singletons the space is Hausdorff. Thus normal implies regular implies Hausdorff and a compact Hausdorff space is normal. Metric spaces are normal because, with $\{i, j\} = \{1, 2\}$, $F_i \subset G_i = \{x : \text{dist}(x, F_i) < \text{dist}(x, F_j)\}$. A famous characterisation of normality is the following.

Lemma 2.61 (Urysohn [21, Chap. 4, Lemma 4]) *A topological space is normal if and only if for every pair of disjoint closed sets F_1 and F_2 there is a continuous function from the space into $[0, 1]$ with $f(F_1) = 0$ and $f(F_2) = 1$.*

This is not to say that $f^{-1}\{0\} = F_1$ and $f^{-1}\{1\} = F_2$. To obtain that level of precision, see Lemma 2.62(b).

A set is a G_δ-set if it is the intersection of a denumerable family of open sets. Closed G_δ-sets feature in Corollary 7.11, Theorems 9.4, 10.2, 10.4, and the following lemma.

Lemma 2.62 (a) *For $A \neq \emptyset$ in a normal topological space X there exists a continuous function $f : X \to [0, 1/2]$ with $f(x) = 0$ for $x \in A$ and $f(x) > 0$ otherwise, if and only if A is a closed G_δ-set.*

(b) *For every disjoint pair, F_1 and F_2, of closed G_δ-sets in a normal space X there is a continuous function $f : X \to [0, 1]$ such that $F_1 = f^{-1}\{0\}$ and $F_2 = f^{-1}\{1\}$.*

Proof (a) For the "only if" part suppose such an f exists. Then $A = f^{-1}\{0\}$ is closed. Also

$$A = f^{-1}\{0\} = \bigcap_{k \in \mathbb{N}} f^{-1}[0, 1/k),$$

and so A is a G_δ-set. Conversely, suppose $A \neq \emptyset$ is closed and $A = \bigcap_{k \in \mathbb{N}} U_k$ where U_k is open. Since $X \setminus U_k$ and A are disjoint closed sets, by Lemma 2.61 (Urysohn), there exists a continuous function $f_k : X \to [0, 1]$ such that $f_k(A) = \{0\}$ and $f_k(X \setminus U_k) = \{1\}$. Now define $f : X \to [0, 1/2]$ by $f(x) = \sum_{k \in \mathbb{N}} 2^{-(k+1)} f_k(x)$. Since the series is uniformly convergent f is continuous, $f(x) = 0$ on A and $f(x) \in (0, 1/2]$ elsewhere.

(b) By Urysohn's lemma there exists a continuous $h_0 : X \to [0, 1]$ with $h_0(F_1) = 0$ and $h_0(F_2) = 1$. Since X is a G_δ-space, by part (a) there exists continuous $h_1 : [0, 1] \to [0, 1/2]$ with $h_1^{-1}\{0\} = F_1$ and $h_2 : [0, 1] \to [1/2, 1]$ with $h_2^{-1}\{1\} = F_2$. Now $f : X \to [0, 1]$ defined by

$$f(x) = \inf_X \left\{ h_2(x), \sup_X \left\{ h_0(x), h_1(x) \right\} \right\}$$

is continuous and has the required property.   □

A topological space is a G_δ-space if all its closed sets are G_δ-sets. Metric spaces are G_δ-spaces because, for every closed set, $F = \bigcap_k G_k$ where $G_k = \{x : \mathrm{dist}\{x, F\} < 1/k\}$, $k \in \mathbb{N}$. However, a G_δ-space need not be normal and a normal space need not be a G_δ-space. In this context, the following characterisation of normality is useful.

Theorem 2.63 *A topological space is normal if and only if for all closed F and open G with $F \subset G$ there exist open sets W_n, $n \in \mathbb{N}$, such that $F \subset \bigcup_n W_n$ and $\overline{W_n} \subset G$ for all $n \in \mathbb{N}$.*

Proof By definition of normality, for given F (closed) $\subset G$ (open) in a normal space there is an open set, W_1 say, with $F \subset W_1 \subset \overline{W_1} \subset G$.

Conversely, let A and B be disjoint closed sets in X. Then, by hypothesis, there exist open sets W_n such that $A \subset \bigcup_n W_n$ and $\overline{W_n} \cap B = \emptyset$. Similarly, there exist open sets V_n such that $B \subset \bigcup_n V_n$ and $\overline{V_n} \cap A = \emptyset$.

Now define open sets G_n, H_m, W and V by

$$G_n = W_n \setminus \bigcup_{i \leq n} \overline{V_i}, \quad H_m = V_m \setminus \bigcup_{j \leq m} \overline{W_j}, \quad W = \bigcup_n G_n, \quad V = \bigcup_m H_m.$$

Then $A \subset W$ and $B \subset V$, and W and V are open. To see they are disjoint suppose $x \in W \cap V$. Then for some m^* and n^*, $x \in G_{n^*} \cap H_{m^*}$. But then

$$x \in \left(W_{n^*} \setminus \bigcup_{i \leq n^*} \overline{V_i} \right) \cap \left(V_{m^*} \setminus \bigcup_{j \leq m^*} \overline{W_j} \right) = \emptyset.$$

Thus V and W are disjoint, and hence the space is normal.   □

Lemma 2.64 *A completely separable, regular topological space X is a normal G_δ-space.*

Proof Let F (closed) $\subset G$ (open) $\subset X$. By regularity, for every $x \in F$ there exist disjoint open sets G'_x and G''_x with $x \in G'_x$, $X \setminus G \subset G''_x$ and $G'_x \cap G''_x = \emptyset$. Therefore

$$x \in G'_x \subset X \setminus G''_x \subset G \text{ which implies that } x \in G'_x \subset \overline{G'_x} \subset G \text{ for all } x \in F.$$

Since there is no loss in assuming that $G'_x \in \mathcal{G}$, the denumerable base for the topology, the normality of X follows from Theorem 2.63.

To see that X is a G_δ-space, let F be closed. Then for every $x \notin F$ there exists open sets G'_x and G''_x with $F \subset G'_x$, $x \in G''_x$, $G'_x \cap G''_x = \emptyset$ and $G''_x \in \mathcal{G}$. Since $F = \bigcap_{x \notin F} G'_x$ and at most countably many G'_x are needed, F is a G_δ-set and so X is a normal G_δ-space.   □

Lemma 2.65 *A locally compact Hausdorff space is regular. A completely separable, locally compact Hausdorff space is a normal G_δ-space.*

Proof For F closed and $x \notin F$, let $x \in G$ where G is open and $\overline{G}$ is compact. Since X is Hausdorff and $x \notin F \bigcap \overline{G}$ which is compact, there are disjoint open sets V, W with $x \in V$ and $F \bigcap \overline{G} \subset W$. Since $\overline{G}$ is compact in a Hausdorff space it is closed, and so $V \bigcap G$ and $W \bigcup (X \setminus \overline{G})$ are open. Therefore, X is regular since $x \in V \bigcap G$, $F \subset W \bigcup (X \setminus \overline{G})$ and $(V \bigcap G) \bigcap (W \bigcup (X \setminus \overline{G})) = \emptyset$. Lemma 2.64 now implies that a completely separable, locally compact Hausdorff space is a normal G_δ-space. □

Chapter 3
L_∞ and Its Dual

Henceforth, $(X, \mathcal{L}, \lambda)$ is a complete σ-finite measure space and $L_\infty(X, \mathcal{L}, \lambda)$ is defined by (2.8); similarly for $(X, \mathcal{B}, \lambda)$ and $L_\infty(X, \mathcal{B}, \lambda)$. The general terminology will be that of Chap. 2.

In notation for finitely additive measures that will be considered in detail in Chap. 4, the representation theorem for the dual space $L_\infty(X, \mathcal{L}, \lambda)^*$, which is analogous to the Riesz Theorem 2.37 for continuous functions, is the following special case of [35, Thm. 2.3], see also [12, Thm. IV.8.16] and the references therein. This result will be referred to as the Yosida–Hewitt representation of $L_\infty^*(X, \mathcal{L}, \lambda)$.

Theorem 3.1 *For every bounded linear functional on $L_\infty(X, \mathcal{L}, \lambda)$ there exists a finitely additive measure (Definition 4.1) ν on $\mathcal{L}$ such that*

$$\nu(N) = 0 \ \text{for all } N \in \mathcal{N}, \tag{3.1a}$$

$$f(u) = \int_X u \, d\nu \, \text{for all } u \in L_\infty(X, \mathcal{L}, \lambda), \tag{3.1b}$$

$$|\nu|(X) = \|f\|_\infty < \infty. \tag{3.1c}$$

Conversely if ν is a finitely additive measure on X with $\nu(N) = 0$ for all $N \in \mathcal{N}$, then f in (3.1b) is a well-defined bounded linear functional on $L_\infty(X, \mathcal{L}, \lambda)$. The set of finitely additive measure on X with $\nu(N) = 0$ for $N \in \mathcal{N}$ will be denoted by $L_\infty^*(X, \mathcal{L}, \lambda)$. (Note that $|\nu|(N) = 0$ if $N \in \mathcal{N}$ and $\nu \in L_\infty^*(X, \mathcal{L}, \lambda)$.)

Once the integral with respect to finitely additive measures in Theorem 3.1 has been defined, the proof in Sect. 6.2 of the theorem is straightforward (indeed, an interested reader might like to have a quick look now). However, there are differences from standard theory, see Remark 2.26, and the development needs to be handled carefully. In the next few chapters, notation and terminology for finitely additive measures,

© The Author(s), under exclusive license to Springer Nature Switzerland AG 2020
J. Toland, *The Dual of $L_\infty(X, \mathcal{L}, \lambda)$, Finitely Additive Measures and Weak Convergence*, SpringerBriefs in Mathematics,
https://doi.org/10.1007/978-3-030-34732-1_3

their existence and basic properties on σ-algebras will be established. However, the following implications of Theorem 3.1 are more or less immediate.

Corollary 3.2 *If* $\{A_k\} \subset \mathcal{L}$ *is a sequence in* $\mathcal{L}$ *with* $\lambda(A_i \cap A_j) = 0$, $i \neq j$, *then* $\chi_{A_k} \rightharpoonup 0$ *as* $k \to \infty$ *in* $L_\infty(X, \mathcal{L}, \lambda)$, *where* χ_{A_k} *denotes the characteristic function of* A_k.

Proof By Theorem 3.1, for any $f \in L_\infty(X, \mathcal{L}, \lambda)^*$ there is a finitely additive measure ν with

$$f(\chi_{A_k}) = \int_X \chi_{A_k}\, d\nu = \nu(A_k) = \nu^+(A_k) - \nu^-(A_k),$$

where $\nu^\pm$ (the positive and negative parts of ν, see (4.1c)) are non-negative finitely additive measures. Since, by finite additivity and (6.2d),

$$0 \leqslant \sum_{k=1}^K \nu^\pm(A_k) = \nu^\pm\left(\bigcup_{k=1}^K A_k\right) \leqslant \nu^\pm(X) < \infty, \quad \text{for all } K \in \mathbb{N},$$

$\nu^\pm(A_k) \to 0$ and the result follows.   □

When combined with the Hahn–Banach theorem, Theorem 3.1 yields a wide variety of finitely additive measures.

Corollary 3.3 *If* (X, ϱ) *is a Hausdorff topological space and* ϱ *is not the discrete topology, there exists a finitely additive measure* $\nu \neq 0$ *which is zero on* $\mathcal{N}$ *and* $\int_X v\, d\nu = 0$ *for all* $v \in L_\infty(X, \mathcal{B}, \lambda)$ *which are continuous on* X.

Proof From the hypothesis and Remark 2.34, elements of $L_\infty(X, \mathcal{B}, \lambda)$ which are continuous on X form a proper closed linear subspace of $L_\infty(X, \mathcal{B}, \lambda)$. The result is then immediate from Theorem 3.1 and Corollary 2.40 of the Hahn–Banach theorem.   □

The existence of finitely additive measures with particular properties will be considered again in Chap. 5.

Corollary 3.4 *Suppose* $\mathcal{B}$ *is the Borel* σ-algebra of a locally compact Hausdorff space (X, ϱ), and $(X, \mathcal{B}, \lambda)$ is a corresponding measure space. Then for every $\nu \in L_\infty^*(X, \mathcal{B}, \lambda)$ there exists a unique, regular, real Borel measure $\tilde{\nu}$ such that

$$\int_X v\, d\nu = \int_X v\, d\tilde{\nu} \text{ for all } v \in C_0(X, \varrho), \quad |\tilde{\nu}|(X) \leqslant |\nu|(X), \qquad (3.2)$$

and $\tilde{\nu} \geqslant 0$ *on Borel sets when* $\nu \geqslant 0$ *on* $\mathcal{L}$.

Proof For $\nu \in L_\infty^*(X, \mathcal{B}, \lambda)$, define $f \in L_\infty(X, \mathcal{B}, \lambda)^*$ by (3.1b). The restriction $\tilde{f}$ of f to $C_0(X, \varrho)$ has $\|\tilde{f}\|_{C_0(X,\varrho)^*} \leqslant \|f\|_{L_\infty(X,\mathcal{B},\lambda)^*} = |\nu|(X) < \infty$, and the result follows from Theorem 2.37 (Riesz).   □

It follows from Corollary 3.3 that the inequality in (3.2) may be strict and, from the Valadier–Hensgen example in Sect. 6.6, that for a given regular real Borel measure $\tilde{\nu}$ there may be uncountably many finitely additive measures ν which satisfy (3.2).

Chapter 4
Finitely Additive Measures

Finitely additive measures are naturally defined on algebras (collections of sets which are closed under complementation and finite unions), but here they are considered on σ-algebras (closed under complementation and countable unions) because $\mathcal{L}$ in Theorem 3.1 is a σ-algebra. Although the terminology "finitely additive measures" may suggest only a slight modification of familiar measure theory, intuition based on experience with σ-additivity can seriously mislead, see (†) in the Preface which is justified in Remark 6.1.

4.1 Definition, Notation and Basic Properties

Definition 4.1 (*Finitely Additive Measures* [35, Sects. 1.2–1.7]) A finitely additive measure ν on $\mathcal{L}$ is a mapping from $\mathcal{L}$ into $\mathbb{R}$ with

$$\nu(\emptyset) = 0 \text{ and } \sup_{A \in \mathcal{L}} |\nu(A)| < \infty;$$

$$\nu(A \bigcup B) = \nu(A) + \nu(B) \text{ for all } A, \ B \in \mathcal{L} \text{ with } A \bigcap B = \emptyset.$$

A finitely additive measure ν is σ-additive if and only if

$$\nu\left(\bigcup_{k \in \mathbb{N}} E_k\right) = \sum_{k \in \mathbb{N}} \nu(E_k) \text{ for all } \{E_k\} \subset \mathcal{L} \text{ with } E_j \bigcap E_k = \emptyset, j \neq k.$$

In the literature, it is usual for ba($\mathcal{L}$) (bounded and additive on $\mathcal{L}$) to denote the set of finitely additive measures on $\mathcal{L}$. Let $\Sigma(\mathcal{L}) \subset$ ba($\mathcal{L}$) denote the σ-additive measures. □

© The Author(s), under exclusive license to Springer Nature Switzerland AG 2020   31
J. Toland, *The Dual of $L_\infty(X, \mathcal{L}, \lambda)$, Finitely Additive Measures and Weak Convergence*, SpringerBriefs in Mathematics,
https://doi.org/10.1007/978-3-030-34732-1_4

Obviously, the set $\Sigma(\mathcal{L})$ coincides with the set of real measures in Sect. 2.4. However $\lambda \in \mathrm{ba}(\mathcal{L})$ if and only if $\lambda(X) < \infty$. Hence $\lambda \notin \Sigma(\mathcal{L})$ when $\lambda(X) = \infty$ although it is σ-additive on $\mathcal{L}$. However, because by hypothesis λ is σ-finite on $\mathcal{L}$, there exists [28, Lemma 6.9] an integrable function f which is positive everywhere on X. Hence $\gamma(E) := \int_E f \, d\lambda$ defines $\gamma \in \Sigma(\mathcal{L})$ which is positive on $\mathcal{L} \setminus \mathcal{N}$ and can be used as a surrogate for λ in arguments involving finitely additive measures, for example, in the proof of Theorem 10.4.

There follows a review of standard theory sufficient for our purposes. For a comprehensive account see [35], [12, Chap. III] or [6, Chap. 4].

Lemma 4.2 *Suppose $\nu \in \mathrm{ba}(\mathcal{L})$.*

(a) Then $\nu \in \Sigma(\mathcal{L})$ if and only if $\nu(A_k) \to 0$ as $k \to \infty$ for any nested sequence $A_{k+1} \subset A_k$, $A_k \in \mathcal{L}$ with $\bigcap_k A_k = \emptyset$.
(b) If $0 \leqslant \nu \leqslant \gamma$ for some $\gamma \in \Sigma(\mathcal{L})$, then $\nu \in \Sigma(\mathcal{L})$.

Proof (a) Suppose $\nu \in \Sigma(\mathcal{L})$ and, for $\{A_k\} \subset \mathcal{L}$ nested with $\bigcap_k A_k = \emptyset$, let $E_k = A_k \setminus A_{k+1} \in \mathcal{L}$. Then $\{E_k\}$ is a sequence of mutually disjoint sets with $\bigcup_k E_k = A_1$. Since $\nu(A_1)$ is finite and ν is σ-additive

$$\nu(A_1) = \sum_{k=1}^\infty \nu(E_k) = \sum_{k=1}^\infty \nu(A_k \setminus A_{k+1}) = \lim_{k\to\infty} \left(\nu(A_1) - \nu(A_k)\right),$$

whence $\lim_{k\to\infty} \nu(A_k) = 0$.

For the converse suppose $\nu \in \mathrm{ba}(\mathcal{L})$ and $\lim_{k\to\infty} \nu(A_k) = 0$ for all nested $\{A_k\} \subset \mathcal{L}$ with $\bigcap_k A_k = \emptyset$. Then for any sequence of mutually disjoint sets $\{E_j\}$ in $\mathcal{L}$ let $A_k = \bigcup_k^\infty E_j$. By hypothesis $\lim_{k\to\infty} \nu(A_k) = 0$ because $\bigcap_k A_k = \emptyset$ and by finite additivity,

$$\sum_{j=1}^{k-1} \nu(E_j) = \nu\left(\bigcup_{j=1}^\infty E_j\right) - \nu(A_k) \to \nu\left(\bigcup_{j=1}^\infty E_j\right) \text{ as } k \to \infty.$$

Hence $\nu \in \Sigma(\mathcal{L})$.

Part (b) is immediate because its hypothesis implies that ν satisfies the criterion for σ-additivity in (a). $\qquad\square$

The next result, which is a simple case of Nikodym's convergence theorem [8, Thm. 4.6.3(i)], depends on the following trivial observation. Suppose

$$a_{jk} \in \mathbb{R}, \ j,k \in \mathbb{N}, \ 0 \leqslant a_{jk} \leqslant a_{j'k'} \leqslant M \text{ when } j \leqslant j', \ k \leqslant k'.$$

Then

$$\lim_{k\to\infty} \lim_{j\to\infty} a_{jk} = \sup_k \sup_j a_{jk} = \sup_j \sup_k a_{jk} = \lim_{j\to\infty} \lim_{k\to\infty} a_{jk}.$$

Lemma 4.3 *Suppose* $\{\gamma_k\} \subset \Sigma(\mathcal{L})$ *and*

$$0 \leqslant \gamma_k(E) \leqslant \gamma_{k+1}(E) \leqslant M < \infty \text{ for all } E \in \mathcal{L}, \ k \in \mathbb{N}.$$

Then $\gamma \in \Sigma(\mathcal{L})$ *where* $\gamma(E) = \lim_{k\to\infty} \gamma_k(E), \ E \in \mathcal{L}.$

Proof It is obvious that γ is finitely additive. To show $\gamma \in \Sigma(\mathcal{L})$, let $\{E_\ell\} \subset \mathcal{L}$ be a sequence of mutually disjoint sets and put $a_{jk} = \sum_{\ell=1}^{j} \gamma_k(E_\ell)$ so that $0 \leqslant a_{jk} \leqslant a_{j'k'} \leqslant M$ if $j \leqslant j', k \leqslant k'$. Now since $\gamma_k \in \Sigma(\mathcal{L})$,

$$\gamma\left(\bigcup_{1}^{\infty} E_\ell\right) = \lim_{k\to\infty} \gamma_k\left(\bigcup_{1}^{\infty} E_\ell\right) = \lim_{k\to\infty} \lim_{j\to\infty} \sum_{\ell=1}^{j} \gamma_k(E_\ell)$$

$$= \sup_{k} \sup_{j} \sum_{\ell=1}^{j} \gamma_k(E_\ell) = \sup_{j} \sup_{k} \sum_{\ell=1}^{j} \gamma_k(E_\ell)$$

$$= \lim_{j\to\infty} \lim_{k\to\infty} \sum_{\ell=1}^{j} \gamma_k(E_\ell) = \lim_{j\to\infty} \sum_{\ell=1}^{j} \gamma(E_\ell) = \sum_{\ell=1}^{\infty} \gamma(E_\ell),$$

which shows $\gamma \in \Sigma(\mathcal{L})$.   □

Finitely additive measures form a real linear space with, for $E \in \mathcal{L}$,

$$(\alpha_1 \nu_1 + \alpha_2 \nu_2)(E) = \alpha_1 \nu_1(E) + \alpha_2 \nu_2(E), \ \nu_i \in \mathrm{ba}(\mathcal{L}), \ \alpha_i \in \mathbb{R}, \ i = 1, 2,$$

and a partial ordering $\nu_1 \leqslant \nu_2$ is defined by

$$\nu_1(E) \leqslant \nu_2(E) \text{ for all } E \in \mathcal{L}.$$

For $\nu_1, \nu_2 \in \mathrm{ba}(\mathcal{L})$ and $E \in \mathcal{L}$ let

$$(\nu_1 \vee \nu_2)(E) = \sup_{\{F\in\mathcal{L}:F\subset E\}} \left\{\nu_1(F) + \nu_2(E \setminus F)\right\},$$

$$(\nu_1 \wedge \nu_2)(E) = \inf_{\{F\in\mathcal{L}:F\subset E\}} \left\{\nu_1(F) + \nu_2(E \setminus F)\right\}, \tag{4.1a}$$

from which it is obvious that

$$\nu_1 \wedge \nu_2 = -\left((-\nu_1) \vee (-\nu_2)\right) \text{ and } \nu_1 \vee \nu_2 = -\left((-\nu_1) \wedge (-\nu_2)\right). \tag{4.1b}$$

Theorem 4.4 *(a)*

$$\nu_1, \nu_2 \in \mathrm{ba}(\mathcal{L}) \Rightarrow \nu_1 \wedge \nu_2, \ \nu_1 \vee \nu_2 \in \mathrm{ba}(\mathcal{L}),$$

$$\nu_1, \nu_2 \in \Sigma(\mathcal{L}) \Rightarrow \nu_1 \wedge \nu_2, \ \nu_1 \vee \nu_2 \in \Sigma(\mathcal{L}).$$

(b) For $\nu \in \mathrm{ba}(\mathcal{L})$, let $\nu^+ = \nu \vee 0$, $\nu^- = (-\nu) \vee 0$. Then

$$\nu^\pm \geqslant 0, \quad \nu = \nu^+ - \nu^- \text{ and } \nu^+ \wedge \nu^- = 0. \tag{4.1c}$$

(c) If $\nu \in \mathrm{ba}(\mathcal{L})$ and γ_1, $\gamma_2 \in \Sigma(\mathcal{L})$ with $\gamma_1 \leqslant \nu \leqslant \gamma_2$, then $\nu \in \Sigma(\mathcal{L})$.
(d) If $\nu \in \mathrm{ba}(\mathcal{L})$ and $\nu(E) = 0$ for all $E \in \mathcal{N}$, then $\nu^\pm(E) = 0$ for all $E \in \mathcal{N}$.

Remark 4.5 When $\nu \in \Sigma(\mathcal{L})$ (equivalently when ν is a real measure) (4.1c) is equivalent to (2.4) in Sect. 2.4 because of the Hahn Decomposition Theorem 2.17. For $\nu \in \mathrm{ba}(\mathcal{L})$, $\nu^\pm$ are referred to as the positive and negative parts of ν, and $|\nu| := \nu^+ + \nu^-$ as the total variation of ν (see Theorem 3.1). $\qquad \square$

Proof (a) Let $E_1, E_2 \in \mathcal{L}$ where $E_1 \bigcap E_2 = \emptyset$ and for $F \subset E_1 \bigcup E_2$ let $F_i = E_i \bigcap F$, $i = 1, 2$. Then by finite additivity,

$$(\nu_1 \vee \nu_2)\left(E_1 \bigcup E_2\right) = \sup_{\{F \in \mathcal{L} : F \subset E_1 \bigcup E_2\}} \left\{ \nu_1(F) + \nu_2\left((E_1 \bigcup E_2) \setminus F\right) \right\}$$

$$= \sup_{\{F \in \mathcal{L} : F \subset E_1 \bigcup E_2\}} \left\{ \nu_1(F_1) + \nu_1(F_2) + \nu_2(E_1 \setminus F_1) + \nu_2(E_2 \setminus F_2) \right\}$$

$$= \sup_{\{F_1 \in \mathcal{L} : F_1 \subset E_1\}} \left\{ \nu_1(F_1) + \nu_2(E_1 \setminus F_1) \right\}$$

$$+ \sup_{\{F_2 \in \mathcal{L} : F_2 \subset E_2\}} \left\{ \nu_1(F_2) + \nu_2(E_2 \setminus F_2) \right\} = (\nu_1 \vee \nu_2)(E_1) + (\nu_1 \vee \nu_2)(E_2).$$

So $\nu_1 \vee \nu_2 \in \mathrm{ba}(\mathcal{L})$, and hence $\nu_1 \wedge \nu_2 = -\left((-\nu_1) \vee (-\nu_2)\right) \in \mathrm{ba}(\mathcal{L})$ by (4.1b).

Now suppose $\nu_1, \nu_2 \in \Sigma(\mathcal{L})$ and $\{E_k\} \subset \mathcal{L}$ is a sequence of mutually disjoint sets. Then, for $\epsilon \in (0, 1)$ and $k \in \mathbb{N}$, there exists $F_k \subset E_k$, $F_k \in \mathcal{L}$, such that

$$\nu_1(F_k) + \nu_2(E_k \setminus F_k) \leqslant (\nu_1 \wedge \nu_2)(E_k) + \epsilon^k.$$

Then, by countable additivity, with $F := \bigcup_k F_k \subset \bigcup_k E_k =: E$,

$$(\nu_1 \wedge \nu_2)(E) \leqslant \nu_1(F) + \nu_2(E \setminus F) = \sum_{k=1}^{\infty} \left(\nu_1(F_k) + \nu_2(E_k \setminus F_k) \right)$$

$$\leqslant \sum_{k=1}^{\infty} \left((\nu_1 \wedge \nu_2)(E_k) + \epsilon^k \right) = \sum_{k=1}^{\infty} (\nu_1 \wedge \nu_2)(E_k) + \frac{\epsilon}{1 - \epsilon},$$

for all $\epsilon \in (0, 1)$. Hence

$$(\nu_1 \wedge \nu_2)\left(\bigcup_{k=1}^{\infty} E_k\right) = (\nu_1 \wedge \nu_2)(E) \leqslant \sum_{k=1}^{\infty} (\nu_1 \wedge \nu_2)(E_k).$$

For the opposite inequality note that, with $F \in \mathcal{L}$ and $F \subset E = \bigcup_k E_k$,

$$\nu_1(F) + \nu_2(E \setminus F) = \sum_{k=1}^{\infty} \left(\nu_1(F \bigcap E_k) + \nu_2(E_k \setminus F) \right) \geqslant \sum_{k=1}^{\infty} (\nu_1 \wedge \nu_2)(E_k),$$

whence

$$(\nu_1 \wedge \nu_2) \left(\bigcup_{k=1}^{\infty} E_k \right) = (\nu_1 \wedge \nu_2)(E) \geqslant \sum_{k=1}^{\infty} (\nu_1 \wedge \nu_2)(E_k).$$

Hence $\nu_1 \wedge \nu_2 \in \Sigma(\mathcal{L})$, and it follows that $\nu_1 \vee \nu_2 = -((-\nu_1) \wedge (-\nu_2)) \in \Sigma(\mathcal{L})$.

(b) By definition $\nu^{\pm} \geqslant 0$ and

$$\inf_{\{F \in \mathcal{L}: F \subset E\}} \nu(F) = -\nu^{-}(E), \qquad \sup_{\{F \in \mathcal{L}: F \subset E\}} \nu(F) = \nu^{+}(E). \tag{4.2}$$

Since $\nu(E) = \nu(F) + \nu(E \setminus F)$ for $E, F \in \mathcal{L}$ with $F \subset E$, it follows that

$$\nu(F) - \nu^{-}(E) \leqslant \nu(E) \leqslant \nu(F) + \nu^{+}(E), \tag{4.3}$$

and $\nu = \nu^{+} - \nu^{-}$ follows by taking the supremum on the left and the infimum on the right of (4.3), over $F \in \mathcal{L}$ with $F \subset E$, using (4.2).

Finally suppose $\nu^{+} \wedge \nu^{-} \neq 0$. Then, since $\nu^{\pm} \geqslant 0$ are finitely additive by part (a) there exists $E \in \mathcal{L}$ and $\alpha > 0$ such that for any $F \subset E$, $F \in \mathcal{L}$,

$$0 < \alpha \leqslant \nu^{+}(E \setminus F) + \nu^{-}(F) = \nu^{+}(E) - \nu^{+}(F) + \nu^{-}(F) = \nu^{+}(E) - \nu(F).$$

Hence for all $F \subset E$, $F \in \mathcal{L}, \nu(F) \leqslant \nu^{+}(E) - \alpha$. Since $\alpha > 0$ this contradicts (4.2) and it follows that $\nu^{+} \wedge \nu^{-} = 0$.

(c) Since $0 \leqslant \nu - \gamma_1 \leqslant \gamma_2 - \gamma_1$ and $\gamma_2 - \gamma_1 \in \Sigma(\mathcal{L})$, Lemma 4.2 implies $\nu - \gamma_1 \in \Sigma(\mathcal{L})$, and hence $\nu \in \Sigma(\mathcal{L})$ as required.

(d) The result follows from the hypothesis and (4.2) since $F \in \mathcal{N}$ when $F \in \mathcal{L}$ and $F \subset E \in \mathcal{N}$, because $(X, \mathcal{L}, \lambda)$ is complete. $\qquad \square$

Definition 4.6 (*Absolute Continuity and Singularity* [6, Chap. 6]) For $\nu_1, \nu_2 \in ba(\mathcal{L})$ write $\nu_1 \ll \nu_2$ (ν_1 is absolutely continuous with respect to ν_2), if for all $\epsilon > 0$ there exists δ such $|\nu_1(E)| < \epsilon$ when $|\nu_2|(E) < \delta$, and write $\nu_1 \perp \nu_2$ (ν_1 and ν_2 are mutually singular) if for every $\epsilon > 0$ there exists $E \in \mathcal{L}$ such that $|\nu_1|(E) + |\nu_2|(X \setminus E) < \epsilon$. $\qquad \square$

Remark 4.7 In the special case when $\nu_1, \nu_2 \in \Sigma(\mathcal{L})$ the above definitions are equivalent to simpler statements [6, Thms. 6.1.6 & 6.1.17]:

$\nu_1 \ll \nu_2$ if and only if $|\nu_2|(E) = 0$ implies $\nu_1(E) = 0$;

$\nu_1 \perp \nu_2$ if and only if $|\nu_1|(E) + |\nu_2|(X \setminus E) = 0$ for some $E \in \mathcal{L}$.

However, it is important to remember that a finitely additive measure ν which vanishes on $\mathcal{N}$ need not satisfy $\nu \ll \lambda$ in Definition 4.6 if $\nu \notin \Sigma(\mathcal{L})$. $\square$

4.2   Purely Finitely Additive Measures

Definition 4.8   (*Purely Finitely Additive Measures*) $\mu \in \mathrm{ba}(\mathcal{L})$ is purely finitely additive if

$$\left\{ \gamma \in \Sigma(\mathcal{L}) : 0 \leqslant \gamma \leqslant \mu^+ \right\} = \left\{ 0 \right\} = \left\{ \gamma \in \Sigma(\mathcal{L}) : 0 \leqslant \gamma \leqslant \mu^- \right\}.$$

$\square$

Clearly, μ is purely finitely additive if and only if μ^+ and μ^- are purely finitely additive and $\Pi(\mathcal{L}) \cap \Sigma(\mathcal{L}) = \{0\}$, where $\Pi(\mathcal{L})$ denotes the set of purely finitely additive measures.

Remark 4.9   When $\mu_1, \mu_2 \in \Pi(\mathcal{L})$ and $\mu_1 \leqslant \nu \leqslant \mu_2$ for some $\nu \in \mathrm{ba}(\mathcal{L})$, it is immediate that $\nu \in \Pi(\mathcal{L})$ since $\nu^+ \leqslant \mu_2^+$, $\nu^- \leqslant \mu_1^-$. Similarly, $|\nu| \leqslant \mu \in \Pi(\mathcal{L})$ implies $\nu \in \Pi(\mathcal{L})$. $\square$

Theorem 4.10   *(a)* $\nu \in \mathrm{ba}(\mathcal{L})$ *is purely finitely additive if and only if*

$$\nu^\pm \wedge \gamma = 0 \text{ for all } 0 \leqslant \gamma \in \Sigma(\mathcal{L}).$$

(b) For $\mu, \mu_1, \mu_2 \in \Pi(\mathcal{L})$ *and* $\alpha \in \mathbb{R}$,

$$\alpha\mu, \quad \mu_1 + \mu_2, \quad |\mu|, \quad \mu_1 \wedge \mu_2, \quad \mu_1 \vee \mu_2 \in \Pi(\mathcal{L}).$$

Proof (a) Suppose $\nu \in \Pi(\mathcal{L})$. Then $0 \leqslant \nu^\pm \wedge \gamma \leqslant \gamma$ for any $0 \leqslant \gamma \in \Sigma(\mathcal{L})$, which implies, by Theorems 4.4(a) and 4.2(b), that $\nu^\pm \wedge \gamma \in \Sigma(\mathcal{L})$ and hence by Definition 4.8 that $\nu^\pm \wedge \gamma = 0$. Conversely, for $\nu \in \mathrm{ba}(\mathcal{L})$ suppose that $\nu^\pm \wedge \gamma = 0$ for all $0 \leqslant \gamma \in \Sigma(\mathcal{L})$. Then $0 \leqslant \gamma \leqslant \nu^\pm$, $\gamma \in \Sigma(\mathcal{L})$, implies that $\gamma = \nu^\pm \wedge \gamma = 0$. Hence $\nu \in \Pi(\mathcal{L})$ by Definition 4.8.

   (b) That $\alpha\mu \in \Pi(\mathcal{L})$, $\alpha \in \mathbb{R}$, is immediate from Definition 4.8 since $(-\mu)^\pm = \mu^\mp$. To show that $\mu_1 + \mu_2 \in \Pi(\mathcal{L})$ consider first the case when μ_1 and μ_2 are nonnegative. By part (a) it suffices to show that $(\mu_1 + \mu_2) \wedge \gamma = 0$ for all $0 \leqslant \gamma \in \Sigma(\mathcal{L})$. In other words, it suffices to show, for all $E \in \mathcal{L}$ and all $0 \leqslant \gamma \in \Sigma(\mathcal{L})$, that

$$\inf_{\{F \in \mathcal{L} : F \subset E\}} \{\mu_1(F) + \mu_2(F) + \gamma(E \setminus F)\} = 0.$$

Since $\mu_i \wedge \gamma = 0$, for any $\epsilon > 0$ there exist $E \supset F_i \in \mathcal{L}$, with

$$\mu_i(F_i) + \gamma(E \setminus F_i) \leqslant \epsilon, \ i = 1, 2.$$

Therefore since $\mu_i \geqslant 0$,

$$\mu_1(F_1 \bigcap F_2) + \mu_2(F_1 \bigcap F_2) + \gamma(E \setminus (F_1 \bigcap F_2))$$
$$\leqslant \mu_1(F_1) + \mu_2(F_2) + \gamma(E \setminus F_1) + \gamma(E \setminus F_2) \leqslant 2\epsilon. \quad (4.4)$$

This proves $\mu_1 + \mu_2 \in \Pi(\mathcal{L})$ for non-negative $\mu_i \in \Pi(\mathcal{L})$ and hence $|\mu| = \mu^+ + \mu^- \in \Pi(\mathcal{L})$ for any $\mu \in \Pi(\mathcal{L})$.

More generally,

$$|\mu_1 + \mu_2| \leqslant \mu_1^+ + \mu_2^+ + \mu_1^- + \mu_2^- \in \Pi(\mathcal{L}) \text{ when } \mu_i \in \Pi(\mathcal{L}), \ i = 1, 2,$$

and $\mu_1 + \mu_2 \in \Pi(\mathcal{L})$ follows by Remark 4.9.

Similarly

$$|\mu_1 \wedge \mu_2| + |\mu_1 \vee \mu_2| \leqslant 2(|\mu_1| + |\mu_2|), \ \mu_1, \mu_2 \in \Pi(\mathcal{L}),$$

implies that $\mu_1 \wedge \mu_2$ and $\mu_1 \vee \mu_2$ are in $\Pi(\mathcal{L})$. $\qquad\square$

By (4.1a) and Theorem 4.10(a), a non-negative $\nu \in \text{ba}(\mathcal{L})$ is purely finitely additive if and only if for every $0 \leqslant \gamma \in \Sigma(\mathcal{L})$, $E \in \mathcal{L}$ and $\epsilon > 0$ there exists

$$F \in \mathcal{L}, \ F \subset E \text{ with } \nu(E \setminus F) + \gamma(F) < \epsilon. \quad (4.5)$$

Since $\mathcal{L}$ is a σ-algebra this remark can be significantly refined.

Lemma 4.11 *For* $0 \leqslant \gamma \in \Sigma(\mathcal{L}), 0 \leqslant \mu \in \Pi(\mathcal{L})$ *and* $\epsilon > 0$ *there exists* $F \in \mathcal{L}$ *with* $\gamma(F) \leqslant \epsilon$ *and* $\mu(F) = \mu(X)$.

Proof If $\mu(X) = 0$ the result holds with $F = \emptyset$.

So suppose $\mu(X) > 0$ and for given $\epsilon > 0$ let $e_k = 2^{-k}\epsilon$. Then with $E_0 = X$, by (4.5) there exists $F_1 \in \mathcal{L}$ with $\mu(E_0 \setminus F_1) + \gamma(F_1) < e_1$. With $E_1 = E_0 \setminus F_1$, in $\mathcal{L}$ there exists $F_2 \subset E_1$, $F_2 \in \mathcal{L}$, with $\mu(E_1 \setminus F_2) + \gamma(F_2) < e_2$. Then with $E_2 = E_1 \setminus F_2$, there exists $F_3 \subset E_2$, $F_3 \in \mathcal{L}$, with $\mu(E_2 \setminus F_3) + \gamma(F_3) < e_3$. Proceeding by induction, for $k \in \mathbb{N}_0$,

$$\mu(E_k \setminus F_{k+1}) + \gamma(F_{k+1}) < e_{k+1}, \ E_k \supset F_{k+1} \in \mathcal{L}, \ E_{k+1} = E_k \setminus F_{k+1}.$$

It follows that for $K \in \mathbb{N}$,

$$0 \leqslant \mu\left(E_0 \setminus \bigcup_{k=1}^{K} F_k\right) \leqslant e_K \text{ and } \gamma\left(\bigcup_{k=1}^{K} F_k\right) \leqslant \sum_{k=1}^{K} e_k < \epsilon.$$

Now since $\mathcal{L}$ is a σ-algebra, $F = \bigcup_{k=1}^{\infty} F_k \in \mathcal{L}$ with $\gamma(F) \leqslant \epsilon$. Moreover, $\mu(X) = \mu(F)$ since $\mu(X \setminus F) \leqslant \epsilon_K$ for all K and μ is finitely additive. This completes the proof.   □

The sense in which a purely finitely additive measure on a σ-algebra is singular with respect to any σ-additive measure is captured by the following.

Theorem 4.12 *Let $0 \leqslant \nu \in \mathrm{ba}(\mathcal{L})$. Then $\nu \in \Pi(\mathcal{L})$ if and only if for every non-negative $\gamma \in \Sigma(\mathcal{L})$ there exists a sequence $\{E_k\} \subset \mathcal{L}$ with*

$$E_{k+1} \subset E_k, \quad \nu(E_k) = \nu(X) \text{ for all } k \text{ and } \gamma(E_k) \to 0 \text{ as } k \to \infty.$$

Proof Suppose $0 \leqslant \nu \in \Pi(\mathcal{L})$ and $0 \leqslant \gamma \in \Sigma(\mathcal{L})$. Then by the preceding lemma, for $n \in \mathbb{N}$ there exists $F_n \in \mathcal{L}$ with $\nu(F_n) = \nu(X)$ and $0 \leqslant \gamma(F_n) < 1/n$. Let $E_k = \bigcap_{n=1}^{k} F_n$. Then $\nu(E_k) = \nu(X)$ because, by finite additivity, $\nu(X \setminus E_k) \leqslant \sum_{n=1}^{k} \nu(X \setminus F_n) = 0$, and $0 \leqslant \gamma(E_k) \leqslant \nu(F_k) \leqslant 1/k$. Conversely, if for every non-negative $\gamma \in \Sigma(\mathcal{L})$ such a sequence $\{E_k\}$ exists, then $\gamma \wedge \nu = 0$. Hence $\nu \in \Pi(\mathcal{L})$ by Theorem 4.10.   □

4.3 Canonical Decomposition: $\mathrm{ba}(\mathcal{L}) = \Sigma(\mathcal{L}) \oplus \Pi(\mathcal{L})$

Theorem 4.13 *Any $\nu \in \mathrm{ba}(\mathcal{L})$ can be written uniquely as $\nu = \mu_\nu + \gamma_\nu$ where $\mu_\nu \in \Pi(\mathcal{L})$, $\gamma_\nu \in \Sigma(\mathcal{L})$. Moreover $\mu_\nu, \gamma_\nu \geqslant 0$ if $\nu \geqslant 0$.*

Proof To show that such a decomposition, if it exists, is unique let $\nu = \mu_i + \gamma_i$, $\mu_i \in \Pi(\mathcal{L})$, $\gamma_i \in \Sigma(\mathcal{L})$, $i = 1, 2$. Then, by Theorem 4.10, $\mu_1 - \mu_2 = \gamma_2 - \gamma_1 \in \Pi(\mathcal{L}) \bigcap \Sigma(\mathcal{L})$ and uniqueness follows since $\Pi(\mathcal{L}) \bigcap \Sigma(\mathcal{L}) = \{0\}$ as noted after Definition 4.8.

Since, by (4.1c), $\nu = \nu^+ - \nu^-$ and, by Theorem 4.10, $\Sigma(\mathcal{L})$ and $\Pi(\mathcal{L})$ are linear spaces, it suffices to show the decomposition exists for non-negative $\nu \in \mathrm{ba}(\mathcal{L})$. Let

$$\varsigma := \sup \left\{ \gamma(X) : 0 \leqslant \gamma \leqslant \nu, \ \gamma \in \Sigma(\mathcal{L}) \right\}$$

and let $\gamma_k \in \Sigma(\mathcal{L})$ be such that $0 \leqslant \gamma_k \leqslant \nu$ and $\lim_{k \to \infty} \gamma_k(X) = \varsigma$.

Replacing γ_k by $\vee_{j=1}^{k} \gamma_j$, by Theorem 4.4(a) there is no loss in assuming that $\gamma_k \leqslant \gamma_{k+1}$ for all $k \in \mathbb{N}$. Consequently $\gamma_\nu(E) := \lim_{k \to \infty} \gamma_k(E) \leqslant \nu(E) \leqslant \nu(X)$ exists for all $E \in \mathcal{L}$, and $\gamma_\nu \in \Sigma(\mathcal{L})$ by Lemma 4.3. Clearly $0 \leqslant \mu_\nu := \nu - \gamma_\nu \in \mathrm{ba}(\mathcal{L})$ and $\gamma_\nu(X) = \varsigma$.

To show that $\mu_\nu \in \Pi(\mathcal{L})$, suppose $0 \leqslant \gamma \leqslant \mu_\nu$ for some $\gamma \in \Sigma(\mathcal{L})$. Then $0 \leqslant \gamma + \gamma_\nu \leqslant \nu$ which implies that $\gamma(X) + \gamma_\nu(X) \leqslant \varsigma = \gamma_\nu(X)$. Hence $\gamma(X) = 0$, which, by Definition 4.8, implies $0 \leqslant \mu_\nu \in \Pi(\mathcal{L})$. Obviously from the construction, γ_ν and μ_ν are non-negative when $\nu \geqslant 0$. This completes the proof.   □

4.4 $L_\infty^*(X, \mathcal{L}, \lambda)$

This chapter closes by specialising briefly to the set $L_\infty^*(X, \mathcal{L}, \lambda)$ of finitely additive measures that feature in Theorem 3.1 (Yosida–Hewitt).

Definition 4.14 ($L_\infty^*(X, \mathcal{L}, \lambda)$) $\nu \in L_\infty^*(X, \mathcal{L}, \lambda)$ if $\nu \in \mathrm{ba}(\mathcal{L})$ and $\nu(E) = 0$ for all $E \in \mathcal{N}$. □

Theorem 4.15 *Suppose $\nu \in L_\infty^*(X, \mathcal{L}, \lambda)$. Then $\nu \in \Pi(\mathcal{L})$ if and only if $\nu^+ \wedge \gamma = \nu^- \wedge \gamma = 0$ for all non-negative $\gamma \in \Sigma(\mathcal{L}) \bigcap L_\infty^*(X, \mathcal{L}, \lambda)$.*

Proof Suppose $\nu \in L_\infty^*(X, \mathcal{L}, \lambda)$ and $\nu^\pm \wedge \gamma = 0$ for all non-negative $\gamma \in \Sigma(\mathcal{L}) \bigcap L_\infty^*(X, \mathcal{L}, \lambda)$. Let $\hat{\gamma} \in \Sigma(\mathcal{L})$ be arbitrary with $0 \leqslant \hat{\gamma} \leqslant \nu^\pm$. Since, by Theorem 4.4(d), $\nu^\pm = 0$ on $\mathcal{N}$ it follows that $\hat{\gamma} \in \Sigma(\mathcal{L}) \bigcap L_\infty^*(X, \mathcal{L}, \lambda)$. Hence, by hypothesis, $\nu^\pm \wedge \hat{\gamma} = 0$ and, by Definition 4.8, $\nu \in \Pi(\mathcal{L})$.

Conversely, if $\nu \in \Pi(\mathcal{L})$, by Definition 4.8 $\nu^\pm \wedge \gamma = 0$ for all $0 \leqslant \gamma \in \Sigma(\mathcal{L}) \bigcap L_\infty^*(X, \mathcal{L}, \lambda)$. □

Theorem 4.16 *Any $\nu \in L_\infty^*(X, \mathcal{L}, \lambda)$ can be written uniquely as*

$$\nu = \mu_\nu + \gamma_\nu \in \left(L_\infty^*(X, \mathcal{L}, \lambda) \bigcap \Pi(\mathcal{L}) \right) \oplus \left(L_\infty^*(X, \mathcal{L}, \lambda) \bigcap \Sigma(\mathcal{L}) \right). \quad (4.6)$$

Proof If $0 \leqslant \nu \in L_\infty^*(X, \mathcal{L}, \lambda)$ in Theorem 4.13, $0 \leqslant \gamma_\nu \leqslant \nu$ implies that $\gamma_\nu(E) = 0$, $E \in \mathcal{N}$. Hence $\gamma_\nu \in L_\infty^*(X, \mathcal{L}, \lambda)$ and since $\mu_\nu = \nu - \gamma_\nu$, so is μ_ν. The result for general $\nu \in \Pi(\mathcal{L})$ follows since $\nu \in L_\infty^*(X, \mathcal{L}, \lambda)$ implies $\nu^\pm \in L_\infty^*(X, \mathcal{L}, \lambda)$, see Theorem 4.4(d). □

Chapter 5
ℭ: 0–1 Finitely Additive Measures

In the notation of Definition 4.14, this chapter concerns the set

$$\mathfrak{G} = \left\{ \omega \in L_\infty^*(X, \mathcal{L}, \lambda) : \omega(X) = 1, \; \omega(E) \in \{0, 1\}, \; E \in \mathcal{L} \right\} \qquad (5.1)$$

of finitely additive measures, each element of which would, by Theorem 3.1, correspond to an element $f \in L_\infty(X, \mathcal{L}, \lambda)^*$ with the property that

$$f(\chi_X) = 1, \; f(\chi_E) \in \{0, 1\}, \; E \in \mathcal{L}; \qquad f(\chi_E) = 0, \; E \in \mathcal{N}.$$

Much of the importance of ℭ in the sequel can be traced back to the next theorem and subsequent remark which suggest that the action of elements of ℭ on $L_\infty(X, \mathcal{L}, \lambda)$ is reminiscent of Dirac measures acting on continuous functions.

Theorem 5.1 *For $u \in L_\infty(X, \mathcal{L}, \lambda)$ and $\omega \in \mathfrak{G}$ there is a unique $\alpha \in I := \big[[u]_\infty^-,$ $[u]_\infty^+\big]$ (see (2.8)) such that*

$$\omega\left(\left\{ x \in X : |u(x) - \alpha| < \epsilon \right\}\right) = 1 \text{ for all } \epsilon > 0. \qquad (5.2)$$

This means (5.2) holds for a unique $\alpha \in \mathbb{R}$, independent of the function chosen to represent the equivalence class $u \in L_\infty(X, \mathcal{L}, \lambda)$.

Proof For $i = 1, 2$ suppose u_i is essentially bounded, $\alpha_i \in \mathbb{R}$ and $u_1(x) = u_2(x)$ for λ-almost all $x \in X$. If (5.2) is satisfied by both $(u_i, \alpha_i), i = 1, 2$, and if $\alpha_1 - \alpha_2 =: 2\epsilon_0 > 0$, then

$$\left\{ x \in X : |u_1(x) - \alpha_1| < \epsilon_0 \right\} \bigcap \left\{ x \in X : |u_2(x) - \alpha_2| < \epsilon_0 \right\} \in \mathcal{N}.$$

Hence by finite additivity the ω-measure of their union would be 2, contradicting $\omega \in \mathfrak{G}$. Thus (5.2) cannot be satisfied by both (u_1, α_1) and (u_2, α_2) simultaneously

J. Toland, *The Dual of $L_\infty(X, \mathcal{L}, \lambda)$, Finitely Additive Measures and Weak Convergence*, SpringerBriefs in Mathematics, https://doi.org/10.1007/978-3-030-34732-1_5

if $\alpha_1 \neq \alpha_2$ when $u_1 = u_2$ λ-almost everywhere. On the other hand if $\alpha_1 = \alpha_2$ and $u_1 = u_2$ λ-almost everywhere, either they both satisfy (5.2) or neither satisfies (5.2). This shows that the value of α (if any) for which (5.2) holds is independent of the function used to represent the equivalence class $u \in L_\infty(X, \mathcal{L}, \lambda)$ and, for given $u \in L_\infty(X, \mathcal{L}, \lambda)$, (5.2) holds for at most one α. By a similar argument $\alpha \in I$ if (5.2) holds.

Now suppose for given u there is no such α. Then for all $\alpha \in I$ there exists $\epsilon_\alpha > 0$ with $\omega(\{x \in X : |u(x) - \alpha| < \epsilon_\alpha\}) = 0$. By compactness, $I \subset \bigcup_{k=1}^K (\alpha_k - \epsilon_{\alpha_k}, \alpha_k + \epsilon_{\alpha_k})$ and consequently

$$\omega(X) = \omega\left(\left\{x : u(x) \in \bigcup_{k=1}^K (\alpha_k - \epsilon_{\alpha_k}, \alpha_k + \epsilon_{\alpha_k})\right\}\right)$$

$$\leqslant \sum_{k=1}^K \omega\left(\left\{x : u(x) \in (\alpha_k - \epsilon_{\alpha_k}, \alpha_k + \epsilon_{\alpha_k})\right\}\right) = 0.$$

Since $\omega(X) = 1$, (5.2) holds for a unique α. □

Remark 5.2 If $\omega \in \mathcal{G}$ but $\omega \notin \Sigma(\mathcal{L})$ (which often happens, see Theorem 5.9 below) it does not follow from (5.2) that $\omega(\{x \in X : u(x) = \alpha\}) = 1$. Nevertheless (5.2) is what is meant by saying that in the sense of finitely additive measures $u \in L_\infty(X, \mathcal{L}, \lambda)$ is a constant ω-almost everywhere. □

The important fact that elements of $\mathcal{G}$ proliferate is a consequence of Zorn's lemma in the light of a well-known one-to-one correspondence between elements of $\mathcal{G}$ and families of sets in $\mathcal{L} \setminus \mathcal{N}$ called ultrafilters in $(X, \mathcal{L}, \lambda)$. This correspondence leads to Theorem 5.6 on the existence of elements of $\mathcal{G}$ with certain properties and ultimately to (†) in the Preface.

5.1 ᵍ and Ultrafilters

Definition 5.3 (*Ultrafilter*) Given $(X, \mathcal{L}, \lambda)$, $\mathcal{F} \subset \mathcal{L}$ is a filter if

(i) $X \in \mathcal{F}$ and $\mathcal{N} \bigcap \mathcal{F} = \emptyset$,

(ii) $E_1, E_2 \in \mathcal{F} \Rightarrow E_1 \bigcap E_2 \in \mathcal{F}$,

(iii) $E_2 \supset E_1 \in \mathcal{F} \Rightarrow E_2 \in \mathcal{F}$.

A filter $\mathcal{U}$ which is maximal with respect to set inclusion, i.e. for a filter $\mathcal{F}$

(iv) $\mathcal{U} \subset \mathcal{F} \Rightarrow \mathcal{F} = \mathcal{U}$, is called an ultrafilter in $(X, \mathcal{L}, \lambda)$. The set of ultrafilters is denoted by $\mathfrak{U}$. $\square$

Theorem 5.4 *(a) For $\mathcal{U} \in \mathfrak{U}$, an element $\omega_{\mathcal{U}} \in \mathfrak{G}$ is defined by*

$$\omega_{\mathcal{U}}(E) := \begin{cases} 1 \text{ if } E \in \mathcal{U} \\ 0 \text{ otherwise} \end{cases}. \tag{5.3a}$$

(b) For $\omega \in \mathfrak{G}$, an element $\mathcal{U}_{\omega} \in \mathfrak{U}$ is defined by

$$\mathcal{U}_{\omega} := \left\{ E \in \mathcal{L} : \omega(E) = 1 \right\}. \tag{5.3b}$$

Note that $\omega_{\mathcal{U}_{\omega}} = \omega$ when $\omega \in \mathfrak{G}$ and $\mathcal{U}_{\omega_{\mathcal{U}}} = \mathcal{U}$ when $\mathcal{U} \in \mathfrak{U}$.

Proof (a) To show that $\omega_{\mathcal{U}}, \mathcal{U} \in \mathfrak{U}$, is well defined on $\mathcal{L}$ it suffices to show that exactly one of E and $E' = X \setminus E$ is in $\mathcal{U}$ for every $E \in \mathcal{L}$. To begin note that at most one of E and E' is in $\mathcal{U}$, for otherwise $E, E' \in \mathcal{U}$ implies $\emptyset = E \cap E' \in \mathcal{U}$, a contradiction since $\emptyset \in \mathcal{N}$.

To see that one of E, E' is in $\mathcal{U}$, suppose $E \notin \mathcal{U}$. Then suppose that $E \cap F \notin \mathcal{N}$ for all $F \in \mathcal{U}$ and let

$$\mathcal{F} = \mathcal{U} \bigcup \left\{ G \in \mathcal{L} : E \cap F \subset G \text{ for some } F \in \mathcal{U} \right\}.$$

Clearly $\mathcal{F}$ is a filter with $\mathcal{U} \subset \mathcal{F}$ and $E \in \mathcal{F} \setminus \mathcal{U}$. Since this contradicts the maximality of $\mathcal{U}$ in (iv) it follows that $E \cap \hat{F} \in \mathcal{N}$ for some $\hat{F} \in \mathcal{U}$.

It now follows that $E' \notin \mathcal{N}$, since $E \cap \hat{F} \in \mathcal{N}$ and $\hat{F} \notin \mathcal{N}$ by (i). Thus $E \notin \mathcal{U}$ implies that $E' \notin \mathcal{N}$, equivalently $N \in \mathcal{N}$ implies $N' \in \mathcal{U}$. In particular, $E \cap \hat{F} \in \mathcal{N}$ implies that $E' \bigcup \hat{F}' \in \mathcal{U}$, and hence by (ii) $E' \cap \hat{F} = (E' \bigcup \hat{F}') \cap \hat{F} \in \mathcal{U}$. Finally $E' \cap \hat{F} \subset E'$ implies that $E' \in \mathcal{U}$ by (iii). This shows that $E \notin \mathcal{U}$ implies $X \setminus E \in \mathcal{U}$, and hence $\omega_{\mathcal{U}}$ is well defined by (5.3a). It remains to check that $\omega \in \mathrm{ba}(\mathcal{L})$.

If $A \cap B = \emptyset$, $A, B \in \mathcal{L}$, it follows from (i) and (ii) that at most one of $\omega_{\mathcal{U}}(A)$ and $\omega_{\mathcal{U}}(B)$ is 1, and if one of them is 1, then $\omega_{\mathcal{U}}(A \bigcup B) = 1$ by (iii). If both are zero, $A' \cap B' \in \mathcal{U}$ by (ii), and hence $\omega_{\mathcal{U}}(A \bigcup B) = 0$. Hence $\omega_{\mathcal{U}} \in \mathrm{ba}(\mathcal{L})$ and it follows that $\omega_{\mathcal{U}} \in \mathfrak{G}$.

(b) Now suppose $\mathcal{U}_{\omega}$ is given by (5.3b) where $\omega \in \mathfrak{G}$. Then $\mathcal{U}_{\omega}$ satisfies (i) and (iii), by (5.1) and finite additivity. Also, for $E_1, E_2 \in \mathcal{U}_{\omega}$,

$$1 = \omega(E_1 \bigcup E_2) = \begin{cases} \omega(E_1 \setminus E_2) + \omega(E_2) = \omega(E_1 \setminus E_2) + 1 \\ \omega(E_2 \setminus E_1) + \omega(E_1) = \omega(E_2 \setminus E_1) + 1 \end{cases}.$$

Hence $\omega(E_1 \setminus E_2) = 0 = \omega(E_2 \setminus E_1)$ which implies $\omega(E_1 \cap E_2) = 1$. So $E_1 \cap E_2 \in \mathcal{U}_{\omega}$ and $\mathcal{U}_{\omega}$ satisfies (ii).

To see that $\mathcal{U}_{\omega}$ satisfies (iv), suppose it does not and let $\widehat{\mathfrak{F}}$ denote the collection of filters which contain $\mathcal{U}_{\omega}$. If $\mathfrak{F} \subset \widehat{\mathfrak{F}}$ is totally ordered by inclusion, $\bigcup_{F \in \mathfrak{F}} F \in \widehat{\mathfrak{F}}$ is an upper bound for $\mathfrak{F}$. Hence, by Zorn's lemma, $\widehat{\mathfrak{F}}$ has a maximal element,

$\widehat{\mathcal{U}} \in \mathfrak{U}$ say. Since, by assumption, $\mathcal{U}_\omega$ is not maximal, there exists $F \in \widehat{\mathcal{U}} \setminus \mathcal{U}_\omega$ with $\omega_{\widehat{\Omega}}(F) = 1$ and $\omega(F) = 0$. Hence $\omega(F') = 1$ which implies that $F' \in \mathcal{U}_\omega$ and hence that $\omega_{\widehat{\Omega}}(F') = 1$ and $\omega_{\widehat{\Omega}}(F) = 0$. This contradiction proves $\mathcal{U}_\omega$ satisfies (iv). $\square$

The observation that $\mathfrak{G}$ is determined entirely by the ultrafilter structure of $\mathcal{L} \setminus \mathcal{N}$ is considered further in Lemma 9.1 and Corollary 9.2. For example, in a locally compact Hausdorff space X, each $\omega \in \mathfrak{G}$ may be identified with a unique point of X just as a Dirac measure is identified with a singleton (see Remark 9.5). However unlike Dirac measures, infinitely many elements of $\mathfrak{G}$ may be identified with the same point and some elements of $\mathfrak{G}$ are identified with the point at infinity.

A simple consequence of Theorem 5.4 is that

$$\omega(A \cap B) = \omega(A)\omega(B) \text{ for all } A, B \in \mathcal{L}, \ \omega \in \mathfrak{G}. \tag{5.4}$$

5.2 $\mathfrak{G}$ and the λ-Finite Intersection Property

A finitely additive measure on $\mathcal{L}$ belongs to $L_\infty^*(X, \mathcal{L}, \lambda)$ if $\nu(N) = 0$ for all $N \in \mathcal{N}$ and Theorem 3.1 combined with the Hahn–Banach theorem yields the existence of a wide variety of elements of $L_\infty^*(X, \mathcal{L}, \lambda)$. The following complementary approach, also based on Zorn's lemma, concerns the existence of elements of $\mathfrak{G}$ which are determined (non-uniquely) by families of sets with the following property.

Definition 5.5 (λ-finite intersection property) $\mathcal{E} \subset \mathcal{L}$ has the λ-finite intersection property if $E_k \in \mathcal{E}, 1 \leqslant k \leqslant K$, implies $\lambda\left(\bigcap_{k=1}^K E_k\right) > 0$. $\square$

Theorem 5.6 If $\mathcal{E}$ has the λ-finite intersection property, there exists $\omega \in \mathfrak{G}$ with $\omega(E) = 1$ for all $E \in \mathcal{E}$.

Proof Consider the collection $\mathfrak{F}$ of filters $\mathcal{F}$ with $\mathcal{E} \subset \mathcal{F}$. Then $\mathfrak{F} \neq \emptyset$ since

$$\left\{A \in \mathcal{L} : A \supset \bigcap_{k=1}^K E_k, \ E_k \in \mathcal{E}, \ K \in \mathbb{N}\right\} \in \mathfrak{F}.$$

Now suppose that $\mathfrak{T} \subset \mathfrak{F}$ is a family of filters which is totally ordered by set inclusion and let $\mathcal{F}^* = \bigcup_{\mathcal{F} \in \mathfrak{T}} \mathcal{F}$. Then $\mathcal{F}^* \in \mathfrak{F}$ is an upper bound for $\mathfrak{T}$. So by Zorn's lemma, $\mathfrak{F}$ has a maximal element $\mathcal{U}$ which is an ultrafilter. By Theorem 5.4 (a), $\omega_\mathcal{U} \in \mathfrak{G}$ satisfies the theorem. $\square$

The next result shows that for given $\mathcal{E}$ there may be uncountably many $\omega \in \mathfrak{G}$ that satisfy Theorem 5.6.

Corollary 5.7 (a) For $A \in \mathcal{L} \setminus \mathcal{N}$ there exists $\omega \in \mathfrak{G}$ with $\omega(A) = 1$.
(b) Suppose, for $j \in \mathbb{N}$, that $F_j \in \mathcal{L} \setminus \mathcal{N}$ and $F_i \cap F_j \in \mathcal{N}$ when $i \neq j$. Let

$$\mathcal{E} = \left\{ E_k : k \in \mathbb{N} \right\} \text{ where } E_k = \bigcup_{j \geq k} F_j.$$

Then for uncountably many distinct $\omega \in \mathfrak{G} \setminus \Sigma(\mathcal{L})$, $\omega(E_k) = 1$ for all k.

Proof (a) This follows from Theorem 5.6 with $\mathcal{E} = \{A\}$.

(b) For $a \in (0, 1)$ let $q_j(a) \in \mathbb{Q}$, $j \in \mathbb{N}$, be such that $q_{j-1}(a) < q_j(a) \nearrow a$ as $j \to \infty$. Let $Q_a = \{q_j(a) : j \in \mathbb{N}\}$ and note that $Q_a \cap Q_b$ is at most finite if $a \neq b$. Now let $N_a = \eta(Q_a)$ where $\eta : \mathbb{Q} \to \mathbb{N}$ is a bijection. So $\{N_a : a \in (0, 1)\}$ is an uncountable collection of infinite subsets of $\mathbb{N}$ with the property that $N_a \cap N_b$ is at most finite if $a \neq b$. In particular, for $a \neq b \in (0, 1)$ there exists $K(a, b) \in \mathbb{N}$ such that $\{j \in N_a \cap N_b : j \geq K(a, b)\} = \emptyset$. Consequently $a \neq b$ implies $F_i \cap F_j \in \mathcal{N}$ for all $i \in N_a, j \in N_b$ with $i, j \geq K(a, b)$.

Now for $a \in (0, 1)$ note that $E_k^a := \bigcup_{k \leq j \in N_a} F_j \in \mathcal{L}$, since $\mathcal{L}$ is a σ-algebra, and let $\mathcal{E}_a = \{E_k^a : k \in \mathbb{N}\}$. Then by Theorem 5.6 there exists $\omega_a \in \mathfrak{G}$ with $\omega_a(E_k^a) = 1$ for all $k \in \mathbb{N}$. However, by construction, $E_k^a \cap E_k^b \in \mathcal{N}$ for all $k \geq K(a, b), a \neq b$. It follows that $\omega_a \neq \omega_b$ since $1 = \omega_a(E_k^a) \neq \omega_b(E_k^a) = 0$ for k sufficiently large. Moreover, for all a, $E_k^a \subset E_k, k \in \mathbb{N}$, which implies $\omega_a(E_k) = 1$ for all $a \in (0, 1)$ and $k \in \mathbb{N}$.

Finally, that $\omega_a \notin \Sigma(\mathcal{L})$ follows from Lemma 4.2(a) since $\bigcap_k E_k^a = \emptyset$ but $E_{k+1}^a \subset E_k^a$ and $\omega(E_k^a) = 1$ for all k. This completes the proof. □

Example 5.8 As in Example 2.14, the space of all bounded sequences of real numbers indexed by $\mathbb{N}$, usually denoted by ℓ_∞, coincides with the space of bounded functions $u : \mathbb{N} \to \mathbb{R}$. With $\ell_\infty(\mathbb{N}) = L_\infty(X, \mathcal{L}, \lambda)$, where $X = \mathbb{N}$, $\mathcal{L} = \wp(\mathbb{N})$ and λ is counting measure, let $\mathcal{E} = \{E_k : k \in \mathbb{N}\}$, where $E_k = k + \mathbb{N}$. Then $\mathcal{E}$ has the λ-finite intersection property and E_k satisfies the hypotheses of Corollary 5.7 with $F_j = \{j\}$. Hence, there exist uncountably many $\omega \in \mathfrak{G}$ with $\omega(E_k) = 1$ for all k. This example underlies Sect. 6.5 on integration of sequences in $\ell_\infty(\mathbb{N})$, and leads to Remark 9.13.

□

Theorem 5.9 *For $\omega \in \mathfrak{G}$*

(a) either $\omega \in \Pi(\mathcal{L})$ or $\omega \in \Sigma(\mathcal{L})$.
(b) $\omega \in \Sigma(\mathcal{L})$ if and only if for an atom $E_\omega \in \mathcal{L}$ (Definition 2.8)

$$\omega(E) = \frac{\lambda(E \cap E_\omega)}{\lambda(E_\omega)} \text{ for all } E \in \mathcal{L}.$$

Proof (a) If this is false, by Theorem 4.13 $\omega = \mu + \gamma$ where $\mu \in \Pi(\mathcal{L})$ and $\gamma \in \Sigma(\mathcal{L})$ are non-negative, $\omega(X) = 1, \mu(X) \in (0, 1)$, and by Theorem 4.12 there exists $\{E_k\} \subset \mathcal{L}$ with $\mu(E_k) = \mu(X)$ for all k and $\gamma(E_k) \to 0$ as $k \to \infty$.

If $\omega(E_k) = 0$ for some k then $0 = \omega(E_k) \geq \mu(E_k) = \mu(X)$, whence $\mu(X) = 0$ which is false. Therefore, since $\omega \in \mathfrak{G}$, $\omega(E_k) = 1$ for all k and it follows that

$$1 = \omega(E_k) = \mu(E_k) + \gamma(E_k) = \mu(X) + \gamma(E_k) \to \mu(X) \text{ as } k \to \infty.$$

Hence $\mu(X) = 1$ which is also false. This proves (a).

(b) Suppose $\omega \in \mathfrak{G} \cap \Sigma(\mathcal{L})$. Then ω is a real measure and both ω and λ are σ-additive with $\omega \ll \lambda$. Since by hypothesis λ is σ-finite, by Theorem 2.28(b) (Radon–Nikodym) there exists $g \in L_1(X, \mathcal{L}, \lambda)$ with

$$\omega(E) = \int_E g \, d\lambda \text{ for all } E \in \mathcal{L}.$$

So g is non-negative λ-almost everywhere on X, $\lambda(\{x \in X : g(x) \geqslant n\}) \to 0$ as $n \to \infty$, and hence

$$\int_{\{x \in X : g(x) \geqslant n\}} g \, d\lambda \to 0 \text{ as } n \to \infty.$$

In particular,

$$\omega\left(\left\{x \in X : g(x) \geqslant n\right\}\right) = \int_{\{x \in X : g(x) \geqslant n\}} g \, d\lambda \to 0 \text{ as } n \to \infty.$$

Now $\omega \in \mathfrak{G}$ implies that for some $N \in \mathbb{N}$, $\omega(\{x \in X : g(x) \geqslant N\}) = 0$ and it follows that $\lambda(\{x \in X : g(x) \geqslant N\}) = 0$. Thus $g \in L_\infty(X, \mathcal{L}, \lambda)$ and by (5.2), for a unique $\alpha \in \mathbb{R}$,

$$\omega\left(\left\{x \in X : |g(x) - \alpha| < \epsilon\right\}\right) = 1 \text{ for all } \epsilon > 0.$$

Since $\omega \in \Sigma(\mathcal{L})$ it follows that $\omega(E_\omega) = 1$ where $E_\omega = \{x \in X : g(x) = \alpha\}$. Moreover $\lambda(E_\omega) > 0$ since $\omega(E_\omega) = 1$ and $\omega \in \mathfrak{G}$. Therefore

$$\omega(E) = \omega(E \cap E_\omega) = \int_{E \cap E_\omega} \alpha \, d\lambda = \alpha\lambda(E \cap E_\omega) \text{ for all } E \in \mathcal{L}.$$

With $\alpha = 1/\lambda(E_\omega)$, E_ω is an atom with the required properties since $\omega \in \mathfrak{G}$.

Conversely, when A is an atom $\omega(E) := \lambda(E \cap A)/\lambda(A)$, $E \in \mathcal{L}$, defines $\omega \in \mathfrak{G} \cap \Sigma(\mathcal{L})$ since λ is σ-additive.   □

Remark 5.10 This shows that all elements of $\mathfrak{G}$ are purely finitely additive when $\mathcal{L}$ has no atoms. In any case if X has infinitely many mutually disjoint subsets with positive λ-measure, by Corollary 5.7(b) there are uncountably many distinct $\omega \in \mathfrak{G}$ which are purely finitely additive. For example, in $\ell_\infty(\mathbb{N})$ $\mathfrak{G}$ has countably many σ-additive elements given by the atoms $\{n\}$, $n \in \mathbb{N}$, and uncountably many $\omega \in \mathfrak{G}$ which are purely finitely additive are given by Corollary 5.7 and Theorem 5.9. Further statements can be made about $\mathfrak{G}$, e.g. Lemmas 7.7, or 9.1 in a locally compact Hausdorff setting.   □

Chapter 6
Integration and Finitely Additive Measures

The integral of $u \in L_\infty(X, \mathcal{L}, \lambda)$ with respect to $\nu \in L_\infty^*(X, \mathcal{L}, \lambda)$ will been defined and sufficient of its properties established to prove the representation Theorem 3.1 (Yosida–Hewitt). Then the special case of integration with respect to $\omega \in \mathfrak{G}$ will be examined and the essential range of $u \in L_\infty(X, \mathcal{L}, \lambda)$ introduced. The chapter ends with an account of the Valadier–Hensgen example. For a more comprehensive account of integration with respect to finitely additive measures, see [6, Chap. 4], [12, Chap. III] and [35].

6.1 The Integral

To begin note that if $u \in L_\infty(X, \mathcal{L}, \lambda)$ and $w(x) = u(x)$ for λ-almost all $x \in X$, it follows that $|\nu|(\{x : w(x) \neq u(x)\}) = 0$ for all $\nu \in L_\infty^*(X, \mathcal{L}, \lambda)$ since $|\nu|(N) = 0$ when $\lambda(N) = 0$. Therefore, by finite additivity the following construction is independent of whether u or w is used to represent an equivalence class in $L_\infty(X, \mathcal{L}, \lambda)$.

Since, by (4.1c), $\nu = \nu^+ - \nu^-$ and by Theorem 4.4(d) $\nu^\pm \in L_\infty^*(X, \mathcal{L}, \lambda)$ when $\nu \in L_\infty^*(X, \mathcal{L}, \lambda)$, it suffices first to define integration with respect to non-negative $\nu \in L_\infty^*(X, \mathcal{L}, \lambda)$ and to extend by linearity.

For $u \in L_\infty(X, \mathcal{L}, \lambda)$ and $p > \|u\|_\infty$, let

$$\mathcal{P}: \quad -p = y_0 < y_1 < \cdots < y_K < y_{K+1} = p, \quad K \in \mathbb{N}, \tag{6.1a}$$

be a partition of $[-p, p]$ and let

$$\underline{S}_\mathcal{P}(u) := \sum_{k=0}^{K} y_k \nu(E_k(u)) \leqslant \sum_{k=0}^{K} y_{k+1} \nu(E_k(u)) =: \overline{S}_\mathcal{P}(u), \tag{6.1b}$$

where $E_k(u) = \{x : y_k \leqslant u(x) < y_{k+1}\}$, $0 \leqslant k < K$. As remarked above, $\underline{S}_\mathcal{P}(u)$ and $\overline{S}_\mathcal{P}(u)$ are independent of which function u is used to represent an element of

J. Toland, *The Dual of $L_\infty(X, \mathcal{L}, \lambda)$, Finitely Additive Measures and Weak Convergence*, SpringerBriefs in Mathematics, https://doi.org/10.1007/978-3-030-34732-1_6

$L_\infty(X, \mathcal{L}, \lambda)$. Since ν is non-negative, finitely additive and $\nu(X) < \infty$,

$$0 \leqslant \overline{S}_{\mathcal{P}}(u) - \underline{S}_{\mathcal{P}}(u) < \rho(\mathcal{P})\nu(X),$$

where $\rho(\mathcal{P}) = \max\{y_{k+1} - y_k : 0 \leqslant k \leqslant K\}$. Furthermore, if a partition $\mathcal{P}$ is augmented by adding an additional point then $\underline{S}_{\mathcal{P}}(u)$ is not decreased and $\overline{S}_{\mathcal{P}}(u)$ is not increased. It follows that over all partitions $\mathcal{P}$, $\inf_{\mathcal{P}} \overline{S}_{\mathcal{P}}(u) = \sup_{\mathcal{P}} \underline{S}_{\mathcal{P}}(u)$. With these observations, the integral of $u \in L_\infty(X, \mathcal{L}, \lambda)$ with respect to $\nu \geqslant 0$ is defined as

$$\inf_{\mathcal{P}} \overline{S}_{\mathcal{P}}(u) = \int_X u \, d\nu = \sup_{\mathcal{P}} \underline{S}_{\mathcal{P}}(u),$$

$$\text{or equivalently } \lim_{\rho(\mathcal{P}) \to 0} \overline{S}_{\mathcal{P}}(u) = \int_X u \, d\nu = \lim_{\rho(\mathcal{P}) \to 0} \underline{S}_{\mathcal{P}}(u). \tag{6.2a}$$

It follows that when $\nu \geqslant 0$ and $u \leqslant v$ λ-almost everywhere on X,

$$\int_X u \, d\nu \leqslant \int_X v \, d\nu. \tag{6.2b}$$

Now the integral of $u \in L_\infty(X, \mathcal{L}, \lambda)$ with respect to general $\nu \in L_\infty^*(X, \mathcal{L}, \lambda)$ can be defined unambiguously by

$$\int_X u \, d\nu := \int_X u \, d\nu^+ - \int_X u \, d\nu^-, \tag{6.2c}$$

from which it follows that

$$\int_X \chi_E \, d\nu = \nu(E), \ E \in \mathcal{L}, \ \nu \in L_\infty^*(X, \mathcal{L}, \lambda). \tag{6.2d}$$

For simplicity of notation write

$$\int_E u \, d\nu := \int_X \chi_E u \, d\nu, \ E \in \mathcal{L}. \tag{6.3a}$$

Note that for $E \in \mathcal{L}$,

$$\left| \int_E u \, d\nu \right| \leqslant \|u\|_\infty |\nu|(E) \text{ and } \int_E u \, d\nu = 0 \text{ if } E \in \mathcal{N}. \tag{6.4}$$

Moreover, when $E = \bigcup_{n=1}^N E_n$, a finite union of sets in $\mathcal{L}$ with $E_i \cap E_j \in \mathcal{N}$ when $i \neq j$,

$$\int_E u \, d\nu = \sum_{n=1}^{N} \int_{E_n} u \, d\nu. \tag{6.5}$$

Since $(-\nu)^{\pm} = \nu^{\mp}$, it follows from (6.2c) that

$$\int_X \alpha u \, d\nu = \alpha \int_X u \, d\nu, \quad u \in L_\infty(X, \mathcal{L}, \lambda), \ \alpha \in \mathbb{R}. \tag{6.6}$$

Therefore, the integral is linear on $L_\infty(X, \mathcal{L}, \lambda)$ if it is shown that

$$\int_X (u + w) \, d\nu = \int_X u \, d\nu + \int_X w \, d\nu, \ u, w \in L_\infty(X, \mathcal{L}, \lambda), \tag{6.7}$$

and by (6.2c) it is enough to consider $\nu \geqslant 0$.

Let $p = 1 + \|u\|_\infty + \|w\|_\infty$ and for $\epsilon > 0$ let $\mathcal{P}$ be a partition of $[-p, p]$ for which $\overline{S}_{\mathcal{P}}(u) - \underline{S}_{\mathcal{P}}(u) < \epsilon$. Then, by (6.5),

$$\int_X (u + w) \, d\nu = \int_{\bigcup_{k=0}^{K} E_k(u)} (u + w) \, d\nu = \sum_{k=0}^{K} \int_{E_k(u)} (u + w) \, d\nu$$

$$\begin{cases} \geqslant \sum_{k=0}^{K} \int_{E_k(u)} (y_k + w) \, d\nu &= \underline{S}_{\mathcal{P}}(u) + \sum_{k=0}^{K} \int_{E_k(u)} w \, d\nu \\ \leqslant \sum_{k=0}^{K} \int_{E_k(u)} (y_{k+1} + w) \, d\nu &= \overline{S}_{\mathcal{P}}(u) + \sum_{k=0}^{K} \int_{E_k(u)} w \, d\nu \end{cases}$$

$$\begin{cases} = \underline{S}_{\mathcal{P}}(u) + \int_X w \, d\nu \geqslant \int_X u \, d\nu + \int_X w \, d\nu - \epsilon \\ = \overline{S}_{\mathcal{P}}(u) + \int_X w \, d\nu \leqslant \int_X u \, d\nu + \int_X w \, d\nu + \epsilon \end{cases} .$$

Since $\epsilon > 0$ was arbitrary, (6.7) follows when $0 \leqslant \nu \in L_\infty^*(X, \mathcal{L}, \lambda)$. The general case is immediate from (6.2c).

Remark 6.1 When $\gamma \in \Sigma(\mathcal{L}) \cap L_\infty^*(X, \mathcal{L}, \lambda)$, it is obvious that integration of $u \in L_\infty(X, \mathcal{L}, \lambda)$ with respect to γ coincides with Lebesgue integration. However, if ν is not σ-additive, familiar results such as the Monotone convergence theorem do not hold for this integral. For example, by Corollary 5.7(b), on the interval $(0, 1)$ there are uncountably many finitely additive measures $\nu \in \mathfrak{G}$ such that $\nu(0, 1 - 1/k] = 0$ and $\nu(1 - 1/k, 1) = 1$ for all $k > 1, k \in \mathbb{N}$. It follows from (6.2a) that for all such ν

$$\int_0^1 1 \, d\nu = 1 \text{ but } \int_0^{1 - \frac{1}{k}} u \, d\nu = 0 \text{ for all } k \in \mathbb{N} \text{ and all } u \in L_\infty(X, \mathcal{L}, \lambda).$$

This establishes (†) in the Preface. □

By Theorem 4.16, $\nu \in L_\infty^*(X, \mathcal{L}, \lambda)$ can be written as $\nu = \mu + \gamma$ where $\mu \in L_\infty^*(X, \mathcal{L}, \lambda) \cap \Pi(\mathcal{L})$ and $\lambda \gg \gamma \in \Sigma(\mathcal{L}) \cap L_\infty^*(X, \mathcal{L}, \lambda)$. Since λ is a σ-finite measure, by Theorem 2.28(b) (Radon–Nikodym) there exists $g \in L_1(X, \mathcal{L}, \lambda)$ with

$$\int_X u \, d\gamma = \int_X u g \, d\lambda \text{ for all } u \in L_\infty(X, \mathcal{L}, \lambda). \tag{6.8}$$

So (4.6) can be rewritten as

$$\nu = \mu + g\lambda, \quad \mu \in \Pi(\mathcal{L}) \bigcap L_\infty^*(X, \mathcal{L}, \lambda), \quad g \in L_1(X, \mathcal{L}, \lambda). \tag{6.9}$$

The relation of (6.9) to Theorem 2.28(a) when λ is Borel measure is examined in Chap. 10.

6.2 Yosida–Hewitt Representation: Proof of Theorem 3.1

Having prepared the ground in Sects. 4.1 and 6.1, the proof of Theorem 3.1, which identifies elements of $L_\infty(X, \mathcal{L}, \lambda)^*$ with elements of $L_\infty^*(X, \mathcal{L}, \lambda)$, and vice versa, is straightforward.

In Sect. 6.1, it was shown that when $\nu \in L_\infty^*(X, \mathcal{L}, \lambda)$ a bounded linear functional $f \in L_\infty(X, \mathcal{L}, \lambda)^*$ is well defined on $L_\infty(X, \mathcal{L}, \lambda)$ by

$$f(u) = \int_X u \, d\nu \text{ and } |f(u)| \leqslant \|u\|_\infty |\nu|(X).$$

Hence $\|f\|_\infty = \sup\{|f(u)| : \|u\|_\infty = 1\} \leqslant |\nu|(X)$. To see that equality holds note that, since $\nu^+ \wedge \nu^- = 0$, by (4.1c) for any $\epsilon > 0$ there exists a set $A \in \mathcal{L}$ with $\nu^+(A) + \nu^-(A') < \epsilon$, where $A' = X \setminus A$. Let $u = \chi_{A'} - \chi_A \in L_\infty(X, \mathcal{L}, \lambda)$. Then $\|u\|_\infty = 1$ and

$$\begin{aligned}
f(u) &= \int_X u \, d\nu = \int_A u \, d\nu^+ + \int_{A'} u \, d\nu^+ - \int_A u \, d\nu^- - \int_{A'} u \, d\nu^- \\
&= -\nu^+(A) + \nu^+(A') + \nu^-(A) - \nu^-(A') \geqslant \nu^+(A') + \nu^-(A) - \epsilon \\
&= \nu^+(X) - \nu^+(A) + \nu^-(X) - \nu^-(A') - \epsilon \geqslant (|\nu|(X) - 2\epsilon)\|u\|_\infty.
\end{aligned}$$

Therefore $\|f\|_\infty \geqslant |\nu|(X)$ and equality follows.

For the converse, given $f \in L_\infty(X, \mathcal{L}, \lambda)^*$ define ν on $\mathcal{L}$ by $\nu(E) = f(\chi_E)$, $E \in \mathcal{L}$. The goal is to show that $\nu \in L_\infty^*(X, \mathcal{L}, \lambda)$ and (3.1b) holds.

First note that $\nu(E) = 0$ if $E \in \mathcal{N}$ because $\|\chi_E\|_\infty = 0$. Moreover, ν is finitely additive because f is linear and $\chi_{E_1 \cup E_2} = \chi_{E_1} + \chi_{E_2}$ when $E_1 \bigcap E_2 = \emptyset$. It remains only to show that (3.1b) holds.

For $u \in L_\infty(X, \mathcal{L}, \lambda)$ let $\mathcal{P}$ in (6.1a) be a partition into $K + 1$ equal intervals, and let $u_K := \sum_{k=0}^K y_k \chi_{E_k(u)} \in L_\infty(X, \mathcal{L}, \lambda)$. Since $u_K \to u$ in $L_\infty(X, \mathcal{L}, \lambda)$ and f is a bounded linear functional

$$f(u) = \lim_{K \to \infty} f(u_K) = \sum_{k=0}^{K} y_k f\left(\chi_{E_k(u)}\right) = \sum_{k=0}^{K} y_k \nu(E_k(u)) \to \int_X u \, d\nu$$

by (6.2a) and (6.2c). Thus (3.1b) holds and the proof is complete. $\qquad \square$

6.3 Integration with Respect to $\omega \in \mathfrak{G}$

Theorem 6.2 *(a) For $u \in L_\infty(X, \mathcal{L}, \lambda)$ and $\omega \in \mathfrak{G}$*

$$\int_X u \, d\omega = \alpha \text{ and } \int_X |u| \, d\omega = |\alpha| \tag{6.10a}$$

where, as in Theorem 5.1, α is such that

$$\omega\left(\left\{x \in X : |u(x) - \alpha| < \epsilon\right\}\right) = 1 \text{ for all } \epsilon > 0.$$

(b) For $u, v \in L_\infty(X, \mathcal{L}, \lambda)$ and $\omega \in \mathfrak{G}$

$$\int_X uv \, d\omega = \left(\int_X u \, d\omega\right)\left(\int_X v \, d\omega\right). \tag{6.10b}$$

(c) If $F : \mathbb{R} \to \mathbb{R}$ is continuous, $u \in L_\infty(X, \mathcal{L}, \lambda)$ and $\omega \in \mathfrak{G}$,

$$\int F(u) \, d\omega = F\left(\int_X u \, d\omega\right). \tag{6.10c}$$

Proof (a) With α given by Theorem 5.1, the first part of (6.10a) follows because

$$\left|\int_X u \, d\omega - \alpha\right| = \left|\int_X (u - \alpha) \, d\omega\right| = \left|\int_{\{x \in X : |u(x) - \alpha| < \epsilon\}} (u - \alpha) \, d\omega\right| < \epsilon$$

for all $\epsilon > 0$. The second part follows because, for all $\epsilon > 0$

$$\omega\left(\left\{x \in X : \big||u(x)| - |\alpha|\big| < \epsilon\right\}\right) \geq \omega\left(\left\{x \in X : |u(x) - \alpha| < \epsilon\right\}\right) = 1.$$

(b) If

$$\int_X u \, d\omega = \alpha \text{ and } \int_X v \, d\omega = \beta,$$

it follows that for all $\epsilon > 0$,

$$\omega\left(\left\{x \in X : |u(x) - \alpha| < \epsilon\right\}\right) = 1 \text{ and } \omega\left(\left\{x \in X : |v(x) - \beta| < \epsilon\right\}\right) = 1$$

and hence, by (5.4),

$$\omega\left(\left\{x \in X : |u(x)v(x) - \alpha\beta| < \epsilon\right\}\right) = 1, \text{ for all } \epsilon > 0.$$

By (6.10a) the proof of (b) is complete.

   (c) It follows from (6.10b) that (6.10c) holds when F is a polynomial, and therefore for continuous F by approximation using Theorem 2.43 (Weierstrass).   □

6.4   Essential Range of $u \in L_\infty(X, \mathcal{L}, \lambda)$

In a general measure space $(X, \mathcal{L}, \lambda)$, the essential range [28, Chap. 3, Ex. 19] of $u \in L_\infty(X, \mathcal{L}, \lambda)$ is defined as

$$\mathcal{R}(u) = \left\{\alpha \in \mathbb{R} : \lambda(\{x : |u(x) - \alpha| < \epsilon\}) > 0 \text{ for all } \epsilon > 0\right\}. \qquad (6.11a)$$

Lemma 6.3 *For $u \in L_\infty(X, \mathcal{L}, \lambda)$,*

$$\mathcal{R}(u) = \left\{\int_X u \, d\omega : \ \omega \in \mathfrak{G}\right\}. \qquad (6.11b)$$

Proof From Theorems 5.6 and 6.2, the right side contains the left. Since $\omega(E) = 1$ implies $\lambda(E) > 0$, by Theorem 6.2 the left contains the right.   □

For another viewpoint on the essential range, see (7.3). In a Borel setting the essential range can be localised, see Sect. 9.3.

6.5   Integrating $u \in \ell_\infty(\mathbb{N})$ with Respect to $\mathfrak{G}$

Recall from Example 5.8 that bounded real sequences $\{u_j\} \in \ell_\infty$ can be identified with the essentially bounded functions $u \in \ell_\infty(\mathbb{N})$ defined by $u(j) = u_j$, $j \in \mathbb{N}$, and there are uncountably many $\hat{\omega} \in \mathfrak{G}$, each with the property that $\hat{\omega}(E_k) = 1$ for all k, where $E_k = k + \mathbb{N}$.

Theorem 6.4 *A sequence $\{u_j\} \in \ell_\infty$ has a subsequence that converges to $\alpha \in \mathbb{R}$ if and only if there exists $\omega \in \mathfrak{G}$ with $\omega(E_k) = 1$ for all $k \in \mathbb{N}$ and $\int_{\mathbb{N}} u \, d\omega = \alpha$.*

Proof Let $\hat{\omega} \in \mathfrak{G}$ have the property that $\hat{\omega}(E_k) = 1$ for all k and suppose $\alpha = \int_{\mathbb{N}} u \, d\hat{\omega}$ where $u \in \ell_\infty(\mathbb{N})$ is defined by $u(j) = u_j$. Then by (5.4) and Theorem 6.2(a)

$$\hat{\omega}\left(\left\{j \in E_k : |u_j - \alpha| < \epsilon\right\}\right) = 1 \text{ for any } k \in \mathbb{N} \text{ and any } \epsilon > 0.$$

Hence, for any $k \in \mathbb{N}$ and $\epsilon > 0$, there exists $j \geqslant k$ such that $|u_j - \alpha| < \epsilon$. In particular, there exists $j_1 \in \mathbb{N}$ such that $|u_{j_1} - \alpha| < 1$ and $j_2 > j_1$ such that $|u_{j_2} - \alpha| < 1/2$. Proceeding by induction, for each $n \in \mathbb{N}$ there exists $j_n > j_{n-1}$ such that $|u_{j_n} - \alpha| < 1/n$. This shows that $\int_{\mathbb{N}} u \, d\hat{\omega} = \alpha$, $\hat{\omega} \in \mathfrak{G}$, implies a subsequence $u_{j_n} \to \alpha$ as $n \to \infty$.

Now suppose $\{u_j\}$ has a subsequence $u_{j_n} \to \alpha$, let $\mathcal{E} = \{\tilde{E}_k : k \in \mathbb{N}\}$ where $\tilde{E}_k = \{j_n : n \geqslant k\}$ and let $\tilde{\omega} \in \mathfrak{G}$ be a finitely additive measure with $\tilde{\omega}(\tilde{E}_k) = 1$ for all $k \in \mathbb{N}$ given by Theorem 5.6. Then for $\epsilon > 0$ there exists $k \in \mathbb{N}$ such that

$$\tilde{E}_k \subset \{j : |u_j - \alpha| < \epsilon\}, \quad \text{and hence } 1 = \tilde{\omega}(\tilde{E}_k) \leqslant \tilde{\omega}(\{j : |u_j - \alpha| < \epsilon\}).$$

By Theorem 6.2 this implies $\int_{\mathbb{N}} u \, d\tilde{\omega} = \alpha$.   □

In Remark 9.13, the preceding observation is related to the essential range at infinity of $u \in \ell_\infty(\mathbb{N})$.

Remark 6.5 For $u \in \ell_\infty$ it is immediate that

$$u_j \to \alpha \text{ as } j \to \infty \Rightarrow \int_{\mathbb{N}} u \, d\omega = \alpha$$

for all $\omega \in \mathfrak{G}$ with $\omega(E_k) = 1$ for all k. Let $S : \ell_\infty(\mathbb{N}) \to \ell_\infty(\mathbb{N})$ be the shift operator defined by $(Su)(j) = u(j+1)$. Then by (6.10b), for $\omega \in \mathfrak{G}$,

$$\int_{\mathbb{N}} u \, S(u) \, d\omega = \left(\int_{\mathbb{N}} u \, d\omega \right) \left(\int_{\mathbb{N}} S(u) \, d\omega \right), \quad u \in \ell_\infty(\mathbb{N}).$$

In particular, when $u(j) = (-1)^j$ it follows that $u \, S(u) \equiv -1$ and hence that

$$\left(\int_{\mathbb{N}} u \, d\omega \right) \left(\int_{\mathbb{N}} S(u) \, d\omega \right) = -1$$

In other words, $u \mapsto \int_{\mathbb{N}} u \, d\omega$, $u \in \ell_\infty(\mathbb{N})$, is not shift invariant, and therefore is not a Banach limit (Definition 2.41 and [31]) for any $\omega \in \mathfrak{G}$.

However, a Banach limit can be constructed as follows. For $u \in \ell_\infty(\mathbb{N})$ let $Tu \in \ell_\infty(\mathbb{N})$ be defined by

$$(Tu)(j) = \frac{1}{j} \sum_{i=1}^{j} u(i), \quad i, j \in \mathbb{N}.$$

Then $T : \ell_\infty(\mathbb{N}) \to \ell_\infty(\mathbb{N})$ is a bounded linear operator with $\|T\| = 1$ and

$$|Tu(j) - T(Su)(j)| = \frac{1}{j} |u(j+1) - u(1)| \to 0 \text{ as } j \to \infty.$$

It follows that if $\omega \in \mathfrak{G}$ and $\omega(E_k)=1$ for all k then $\omega\{j : |Tu(j) - \alpha| < \epsilon\}=1$ for all $\epsilon > 0$ if and only if $\omega\{j : |T(Su)(j) - \alpha| < \epsilon\} = 1$ for all $\epsilon > 0$. Hence, by Theorem 6.2(a),

$$\int_{\mathbb{N}} Tu \, d\omega = \int_{\mathbb{N}} T(Su) \, d\omega \text{ if } \omega \in \mathfrak{G} \text{ and } \omega(E_k) = 1 \text{ for all } k.$$

Since $u(j) \to \alpha$ implies $Tu(j) \to \alpha$ as $j \to \infty$, it follows that $u \mapsto \int_{\mathbb{N}} Tu \, d\omega$, $u \in \ell_\infty(\mathbb{N})$, is a Banach limit.   □

6.6 The Valadier–Hensgen Example

Independently, Valadier [33] and Hensgen [19] made the following observation which contradicts a claim in [35].

Theorem 6.6 *When $\mathcal{L}$ and λ refer to Lebesgue measure on $X = [0, 1]$, there exists $\mu \in \Pi(\mathcal{L}) \bigcap L_\infty^*(X, \mathcal{L}, \lambda)$ such that*

$$\int_0^1 v \, d\mu = \int_0^1 v \, d\lambda \text{ for all continuous } v : [0, 1] \to \mathbb{R}. \tag{6.12}$$

Proof For $x \in [0, 1]$ let $\mu^x \in \mathfrak{G}$ be given by Theorem 5.6 with the property that $\mu^x\big((x - \epsilon, x + \epsilon) \bigcap [0, 1]\big) = 1$ for all $\epsilon > 0$, and note from Theorem 6.2 that

$$\int_0^1 v \, d\mu^x = v(x) \text{ for all } v \in C[0, 1].$$

For $n \in \mathbb{N}$, let $x_j^n = j/2^n, 0 \leqslant j \leqslant 2^n, j \in \mathbb{N}_0$, and for $v \in C[0, 1]$ let

$$s_n^v = \frac{1}{2^n + 1} \sum_{j=0}^{2^n} v(x_j^n) = \int_0^1 v \, d\mu_n, \text{ where } \mu_n = \frac{1}{2^n + 1} \sum_{j=0}^{2^n} \mu^{x_j^n}. \tag{6.13}$$

For all $n \in \mathbb{N}$, $\mu_n \in L_\infty^*(X, \mathcal{L}, \lambda)$ and $\mu_n(E) \in [0, 1]$, $E \in \mathcal{L}$, since each $\mu^{x_j^n} \in \mathfrak{G}$. Moreover, s_n^v is an nth Riemann sum of $v \in C[0, 1]$ corresponding to partitioning $[0, 1]$ with $2^n + 1$ equally spaced points. Hence $s^v := \{s_n^v\} \in \ell_\infty$ is convergent and

$$\lim_{n \to \infty} s_n^v = \int_0^1 v \, d\lambda, \quad v \in C[0, 1],$$

or, in the notation of Definition 2.41,

$$s^v \in c(\mathbb{N}) \text{ and } l(s^v) = \int_0^1 v \, d\lambda \text{ for all } v \in C[0,1].$$

Now define μ on $\mathcal{L}$ by

$$\mu(E) = L(\{\mu_n(E)\}), \quad E \in \mathcal{L}, \tag{6.14}$$

where $L : \ell_\infty(\mathbb{N}) \to \mathbb{R}$ is a Banach limit, Definition 2.41. Clearly $\mu \geqslant 0$ is finitely additive. Also $\mu(E) = 0$ for all $E \in \mathcal{N}$ since μ_n is zero on $\mathcal{N}$.

To see μ is purely finitely additive, for $k \in \mathbb{N}$ let

$$E_k = [0,1] \cap \left(\bigcup_{\substack{\{j,n:\\ 0 \leqslant j \leqslant 2^n\\ n \in \mathbb{N}\}}} \left(x_j^n - \frac{1}{2^{2n}k}, x_j^n + \frac{1}{2^{2n}k} \right) \right).$$

Note that since $\mu_n(E_k) = 1$ for all $n \in \mathbb{N}$, it follows that $\mu(E_k) = 1 = \mu(X)$ for all k. Moreover, since $\{E_k\}$ is a nested sequence of open subsets of $[0,1]$ with $\lambda(E_k) \leqslant 2/k$, it follows that $\lambda(\bigcap_k E_k) = 0$. Hence $\gamma(E_k) \to \gamma(\bigcap_k E_k) = 0$ if $0 \leqslant \gamma \in \Sigma(\mathcal{L})$. Since $\mu \geqslant 0$ this shows that $\mu \wedge \gamma = 0$ for any $0 \leqslant \gamma \in \Sigma(\mathcal{L}) \cap L_\infty^*(X, \mathcal{L}, \lambda)$ whence, by Theorem 4.15, μ is purely finitely additive. It remains to prove (6.12).

When g is a simple function on $[0,1]$ (i.e. $g = \sum_{m=1}^M g_m \chi_{G_m}$, $g_m \in \mathbb{R}$, $G_m \in \mathcal{L}$) it follows from (6.2d) and (6.14) that

$$L\left(\left\{\int_0^1 g \, d\mu_n\right\}\right) = \int_0^1 g \, d\mu. \tag{6.15}$$

Now for an arbitrary $u \in L_\infty([0,1], \mathcal{L}, \lambda)$ let $u_k \to u$ uniformly on $[0,1]$ where $\{u_k\}$ is a sequence of simple functions. Then for every $k \in \mathbb{N}$

$$\left| L\left(\left\{\int_0^1 u \, d\mu_n\right\}\right) - \int_0^1 u \, d\mu \right| \leqslant \left| L\left(\left\{\int_0^1 u \, d\mu_n - \int_0^1 u_k \, d\mu_n\right\}\right) \right|$$
$$+ \left| L\left(\left\{\int_0^1 u_k \, d\mu_n\right\}\right) - \int_0^1 u_k \, d\mu \right| + \left| \int_0^1 u_k \, d\mu - \int_0^1 u \, d\mu \right|.$$

Now the first and third terms on the right converge to 0 as $k \to \infty$ and by (6.15) the middle term is zero for all k. Hence,

$$L\left(\left\{\int_0^1 u \, d\mu_n\right\}\right) = \int_0^1 u \, d\mu \text{ for all } u \in L_\infty([0,1], \mathcal{L}, \lambda).$$

In particular, for $v \in C[0,1]$

$$\int_0^1 v \, d\mu = L\left(\left\{\int_0^1 v \, d\mu_n\right\}\right) = \lim_{n \to \infty} s_n^v = \int_0^1 v \, d\lambda.$$

Hence (6.12) is proved.   □

Remark 6.7 Yosida and Hewitt [35, Thm. 3.4] claimed that

$$\int_0^1 v\, d\nu = 0 \text{ for all } v \in C[0,1] \tag{6.16}$$

implies ν is purely finitely additive but omitted the proof because it was considered straightforward. However, Valadier [33], and later Hensgen [19], independently showed that this claim is wrong. In Theorem 6.6 $\mu \in \Pi(\mathcal{L})$, $\lambda \in \Sigma(\mathcal{L})$ and so, by the uniqueness statement in Theorem 4.13, $\nu := \mu - \lambda \in L_\infty^*(X, \mathcal{L}, \lambda)$ is neither purely finitely additive nor σ-additive yet satisfies (6.16).

Different choices of the points x_j^n lead to different μ. So if $\{y_j^n : j \in \mathbb{N}\} \bigcap \{x_j^n : j \in \mathbb{N}\} = \emptyset$, there are two measures $\mu_1, \mu_2 \in \Pi(\mathcal{L})$ which satisfy (6.12). Hence $0 \neq \mu_1 - \mu_2 = \nu \in \Pi(\mathcal{L})$ also satisfies (6.16). For further discussion of these issues, see Sect. 10.2.   □

Chapter 7
Topology on $\mathfrak{G}$

7.1 The Space $(\mathfrak{G}, \tau)$

For $A \in \mathcal{L}$, let $\Delta_A = \{\omega \in \mathfrak{G} : \omega(A) = 1\}$ and note that $\Delta_A \subset \Delta_B$ if $A \subset B$ and $\mathfrak{G} \setminus \Delta_A = \Delta_{X \setminus A}$. More generally, from finite additivity it follows that for finite families,

$$\bigcup_1^K \Delta_{A_k} = \Delta_{\bigcup_1^K A_k} \text{ and } \bigcap_1^K \Delta_{A_k} = \Delta_{\bigcap_1^K A_k}, \quad A_k \in \mathcal{L}, \quad (7.1a)$$

whereas for arbitrary families only inclusions hold:

$$\bigcup_\alpha \Delta_{A_\alpha} \subset \Delta_{\bigcup_\alpha A_\alpha} \text{ and } \Delta_{\bigcap_\alpha A_\alpha} \subset \bigcap_\alpha \Delta_{A_\alpha}. \quad (7.1b)$$

For further developments, see Theorem 7.10 and Corollary 7.11.

Lemma 7.1 *Suppose $A, B \in \mathcal{L}$. Then $\Delta_A \cap \Delta_B \neq \emptyset$ if and only if $A \cap B \notin \mathcal{N}$. Moreover, $B \setminus A \notin \mathcal{N}$ if and only if $\Delta_B \not\subset \Delta_A$. Consequently, $\Delta_A = \Delta_B$ if and only if the symmetric difference $(A \setminus B) \cup (B \setminus A) \in \mathcal{N}$.*

Proof The first part is immediate from (7.1a), since elements of $\mathfrak{G}$ are 0 on $\mathcal{N}$, and from Theorem 5.6 since $A \cap B \notin \mathcal{N}$ implies $\Delta_{A \cap B} \neq \emptyset$. It follows that $B \setminus A \notin \mathcal{N}$ if and only if $\Delta_B \setminus \Delta_A \neq \emptyset$ and the observation on the symmetric difference is immediate. $\qquad \square$

Definition 7.2 $((\mathfrak{G}, \tau))$ Let $\mathcal{T} := \{\Delta_A : A \in \mathcal{L}\}$ be a sub-base for the topology τ on $\mathfrak{G}$. (See Remark 2.57.) Equivalently, by (7.1a), $\mathcal{T}$ is a base for τ, see Definition 2.56. Note that Δ_A, $A \in \mathcal{L}$, is open and closed because $\mathfrak{G} \setminus \Delta_A = \Delta_{X \setminus A}$. $\qquad \square$

Theorem 7.3 $(\mathfrak{G}, \tau)$ *is a compact Hausdorff space.*

© The Author(s), under exclusive license to Springer Nature Switzerland AG 2020
J. Toland, *The Dual of $L_\infty(X, \mathcal{L}, \lambda)$, Finitely Additive Measures
and Weak Convergence*, SpringerBriefs in Mathematics,
https://doi.org/10.1007/978-3-030-34732-1_7

Proof Let τ be the topology defined in Definition 7.2. To see that $(\mathfrak{G}, \tau)$ is Hausdorff suppose that $\omega_1 \neq \omega_2$. Then there exists $A \in \mathcal{L}$ such that $\omega_1(A) = 1$ and $\omega_2(A) = 0$. Hence $\omega_1 \in \Delta_A$ and $\omega_2 \in \Delta_{X \setminus A}$ where $\Delta_A \bigcap \Delta_{X \setminus A} = \emptyset$. Thus $(\mathfrak{G}, \tau)$ is Hausdorff.

To show that $(\mathfrak{G}, \tau)$ is compact suppose that $\mathfrak{G} = \bigcup_{A \in \mathcal{A}} \Delta_A$ for some $\mathcal{A} \subset \mathcal{L}$ and that $\mathfrak{G} \neq \bigcup_{A \in \widehat{\mathcal{A}}} \Delta_A$ for any finite subset $\widehat{\mathcal{A}}$ of $\mathcal{A}$. Since $\Delta_{X \setminus A} = \mathfrak{G} \setminus \Delta_A$, it follows that $\bigcap_{A \in \mathcal{A}} \Delta_{X \setminus A} = \emptyset$ but, by (7.1a),

$$\Delta_{(\bigcap_{A \in \widehat{\mathcal{A}}} X \setminus A)} = \bigcap_{A \in \widehat{\mathcal{A}}} \Delta_{X \setminus A} \neq \emptyset \text{ when } \widehat{\mathcal{A}} \subset \mathcal{A} \text{ is finite.}$$

By Lemma 7.1 this implies that $\bigcap_{A \in \widehat{\mathcal{A}}}(X \setminus A) \notin \mathcal{N}$ for any finite $\widehat{\mathcal{A}} \subset \mathcal{A}$.

Since $\mathcal{E} := \{X \setminus A : A \in \mathcal{A}\}$ satisfies the hypotheses of Theorem 5.6, there exists $\omega \in \mathfrak{G}$ such that $\omega(X \setminus A) = 1$, and consequently $\omega(A) = 0$, for all $A \in \mathcal{A}$. Since $\omega \in \mathfrak{G}$ and $\omega \notin \bigcup_{A \in \mathcal{A}} \Delta_A$, this is a contradiction. It follows that $\mathfrak{G} \subset \bigcup_{A \in \widehat{\mathcal{A}}} \Delta_A$ for some finite $\widehat{\mathcal{A}} \subset \mathcal{A}$ and hence, since $\mathcal{T}$ is a base for τ, $(\mathfrak{G}, \tau)$ is compact. $\qquad\square$

7.2 $L_\infty(X, \mathcal{L}, \lambda)$ and $C(\mathfrak{G}, \tau)$ Isometrically Isomorphic

The next observation is that $L_\infty(X, \mathcal{L}, \lambda)$ and $C(\mathfrak{G}, \tau)$ are isometrically isomorphic with the consequence that their duals are isometrically isomorphic, even though one is represented by finitely additive measures, Theorem 3.1, and the other by σ-additive measures, Theorem 2.37 (Riesz).

For $u \in L_\infty(X, \mathcal{L}, \lambda)$ let $L[u] : \mathfrak{G} \to \mathbb{R}$ be defined by

$$L[u](\omega) = \int_X u \, d\omega \text{ for all } \omega \in \mathfrak{G}. \tag{7.2}$$

Note from Lemma 6.3 that the essential range of $u \in L_\infty(X, \mathcal{L}, \lambda)$ is the range of $L[u]$ on $\mathfrak{G}$:

$$\mathcal{R}(u) = L[u](\mathfrak{G}). \tag{7.3}$$

Theorem 7.4 *$L[u]$ is continuous on $(\mathfrak{G}, \tau)$,*

$$\|u\|_\infty = \|L[u]\|_{C(\mathfrak{G}, \tau)} =: \sup_{\omega \in \mathfrak{G}} |L[u](\omega)|, \tag{7.4}$$

and the mapping $u \mapsto L[u]$, from $L_\infty(X, \mathcal{L}, \lambda)$ to $C(\mathfrak{G}, \tau)$ is linear. Moreover, for $u, v \in L_\infty(X, \mathcal{L}, \lambda)$,

$$L[u](\omega)L[v](\omega) = L[uv](\omega) \text{ for all } \omega \in \mathfrak{G}. \tag{7.5}$$

Conversely, for every real-valued continuous function $\mathfrak{u}$ on $(\mathfrak{G}, \tau)$ there exists $u \in L_\infty(X, \mathcal{L}, \lambda)$ with $\mathfrak{u} = L[u]$. In other words, L is an isometric isomorphism between the Banach algebras $L_\infty(X, \mathcal{L}, \lambda)$ and $C(\mathfrak{G}, \tau)$.

Proof For $u_0 \in L_\infty(X, \mathcal{L}, \lambda)$ and $\omega_0 \in \mathfrak{G}$ let $\alpha_0 = L[u_0](\omega_0) = \int_X u_0 \, d\omega_0$ and for $\epsilon_0 > 0$ put $A_0 = \{x \in X : |u_0(x) - \alpha_0| < \epsilon_0/2\}$. Then $\omega_0 \in \Delta_{A_0}$ by Theorem 6.2 and $\{x \in X : |u_0(x) - \hat\alpha| < \epsilon_0/2\} \cap A_0 \in \mathcal{N}$ if $|\hat\alpha - \alpha_0| \geq \epsilon_0$. Therefore $\omega(\{x \in X : |u_0(x) - \hat\alpha| < \epsilon_0/2\}) = 0$, and consequently $\int_X u_0 \, d\omega \neq \hat\alpha$, if $\omega \in \Delta_{A_0}$ and $|\hat\alpha - \alpha_0| \geq \epsilon_0$. It follows that

$$\left| L[u_0](\omega) - \alpha_0 \right| = \left| \int_X u_0 \, d\omega - \alpha_0 \right| < \epsilon_0 \text{ if } \omega \in \Delta_{A_0}.$$

Hence, for any $u_0 \in L_\infty(X, \mathcal{L}, \lambda)$, $L[u_0] : (\mathfrak{G}, \tau) \to \mathbb{R}$ is continuous at any $\omega_0 \in \mathfrak{G}$, as required.

Identity (7.5) is a restatement of (6.10b) in the present notation.

It follows from (6.4), (6.6) and (6.7) that $u \mapsto L[u]$ is linear from $L_\infty(X, \mathcal{L}, \lambda)$ to $C(\mathfrak{G}, \tau)$ and $\sup_{\omega \in \mathfrak{G}} |L[u](\omega)| \leq \|u\|_\infty$. To prove equality, suppose $\|u\|_\infty = [u]^+$ (see (2.8), if not replace u with $-u$). Then for $\epsilon > 0$ let $D_\epsilon = \{x : \|u\|_\infty - u(x) < \epsilon\} \notin \mathcal{N}$. By Theorem 5.6 there exists $\omega \in \mathfrak{G}$ such that $\omega(D_\epsilon) = 1$. Hence, for any $\epsilon > 0$

$$\|u\|_\infty - L[u](\omega) = \|u\|_\infty - \int_X u \, d\omega = \int_{D_\epsilon} (\|u\|_\infty - u) \, d\omega \leq \epsilon.$$

Therefore (7.4) holds.

Now let $\mathcal{L} = \left\{ L[u] : u \in L_\infty(X, \mathcal{L}, \lambda) \right\} \subset C(\mathfrak{G}, \tau)$. Then $L[1] \in \mathcal{L}$ is the constant function 1 and by (7.5) the product of two elements of $\mathcal{L}$ is in $\mathcal{L}$. If $\omega_1 \neq \omega_2$ there is a set $E \in \mathcal{L}$ with $\omega_1(E) = 1$ and $\omega_2(E) = 0$ and, since $\int_X \chi_E \, d\omega_1 = 1 \neq 0 = \int_X \chi_E \, d\omega_2$, $\mathcal{L}$ separates points in $\mathfrak{G}$. Moreover, by (7.4) $\mathcal{L}$ is closed since $L_\infty(X, \mathcal{L}, \lambda)$ is a Banach space. Therefore, by Theorem 2.42 (Stone–Weierstrass); $\mathcal{L} = C(\mathfrak{G}, \tau)$ since $(\mathfrak{G}, \tau)$ is a compact Hausdorff space. This completes the proof. □

7.3 Properties of $\mathfrak{G}$ and τ

Lemma 7.5 $\mathcal{H}$ *is open and closed in* $(\mathfrak{G}, \tau)$ *if and only if* $\mathcal{H} = \Delta_A$, $A \in \mathcal{L}$.

Proof First recall that Δ_A is open and closed because Δ_A and its complement $\mathfrak{G} \setminus \Delta_A = \Delta_{X \setminus A}$ are open. Conversely, if $\mathcal{H}$ is closed in $(\mathfrak{G}, \tau)$ it is compact. Hence, if it is also open, there exists $A_k \in \mathcal{L}$, $1 \leq k \leq K$, such that, by (7.1a),

$$\mathcal{H} = \bigcup_{k=1}^{K} \Delta_{A_k} = \Delta_A \text{ where } A = \bigcup_{k=1}^{K} A_k \in \mathcal{L},$$

and the result follows. □

Theorem 7.6 $(\mathfrak{G}, \tau)$ *is totally disconnected, meaning all its connected sets are singletons.*

Proof Suppose $\mathcal{H} \subset \mathfrak{G}$ and $\omega_1, \omega_2 \in \mathcal{H}$ are distinct. Then, for some $A \in \mathcal{L}, \omega_1(A)=1$ and $\omega_2(A) = 0$, whence $\omega_1 \in \Delta_A$ and $\omega_2 \in \mathfrak{G} \setminus \Delta_A = \Delta_{X \setminus A}$. Therefore, $\mathcal{H}$ is not connected since $\Delta_{X \setminus A}$ and Δ_A are disjoint open sets the union of which is $\mathfrak{G}$. Clearly a singleton $\{\omega\}$ is connected. $\qquad\square$

Lemma 7.7 *For $A \in \mathcal{L}$, Δ_A is a singleton if and only if A is an atom in $(X, \mathcal{L}, \lambda)$.*

Proof If A is not an atom, $A = A_1 \bigcup A_2$ where $A_1 \bigcap A_2 \in \mathcal{N}$, $A_1, A_2 \notin \mathcal{N}$. Hence $\Delta_A = \Delta_{A_1} \bigcup \Delta_{A_2}, \Delta_{A_1} \bigcap \Delta_{A_2} = \emptyset$ and $\Delta_{A_1}, \Delta_{A_2} \neq \emptyset$, and so Δ_A is not a singleton. Conversely, $\omega_1, \omega_2 \in \Delta_A$, $\omega_1 \neq \omega_2$, implies there exist $A_1, A_2 \subset A$ with $\omega_1(A_1) = 1 = \omega_2(A_2)$, $\omega_1(A_2) = 0 = \omega_2(A_1)$. It follows from Lemma 7.1 that $A_1, A_2, (A_1 \setminus A_2) \bigcup (A_2 \setminus A_1) \notin \mathcal{N}$, and so A is not an atom if Δ_A is not a singleton. $\qquad\square$

Corollary 7.8 *When $(X, \mathcal{L}, \lambda)$ has no atoms every element of $\mathfrak{G}$ is purely finitely additive and no element of $\mathfrak{G}$ is isolated.*

Proof Since by definition ω is isolated in $(\mathfrak{G}, \tau)$ if and only if $\{\omega\} = \Delta_A$ for some $A \in \mathcal{L}$, by Lemma 7.7 no element of $\mathfrak{G}$ is isolated if there are no atoms. The pure finite additivity of $\omega \in \mathfrak{G}$ follows by Theorem 5.9. $\qquad\square$

In a Borel measure space $(X, \mathcal{B}, \lambda)$, a further implication of the absence of atoms is the following. Recall from Sect. 2.10 that a topological space (X, ϱ) is completely separable and it has a denumerable basis (a denumerable collection $\mathcal{T} = \{T_i\}$ of non-empty open sets such that every open set is a union of elements of $\mathcal{T}$).

Lemma 7.9 *If $\mathcal{B}$ has no atoms, $(\mathfrak{G}, \tau)$ is not completely separable.*

Proof Let $\{T_i : i \in \mathbb{N}\}$ be an arbitrary denumerable family of non-empty sets in τ and let $\emptyset \neq \Delta_{G_i} \subset T_i$, $G_i \in \mathcal{L}$. Since $\lambda(G_i) > 0$ for all i, by Theorem 2.25 there exists $A \in \mathcal{L}$ which splits $\mathcal{G} = \{G_i : i \in \mathbb{N}\}$. Since $G_i \setminus A \notin \mathcal{N}$ it follows from Lemma 7.1 that $\Delta_{G_i} \not\subset \Delta_A$, and hence that $T_i \not\subset \Delta_A$, for all $i \in \mathbb{N}$. Since $\Delta_A \in \tau$, it follows that $\{T_i : i \in \mathbb{N}\}$ is not a base for τ. This completes the proof. $\qquad\square$

This discussion of $(\mathfrak{G}, \tau)$ is continued in Sect. 9.1, but for that the following refinement of (7.1b) is needed.

Theorem 7.10 *(i) If $\Gamma = \bigcap_{k \in \mathbb{N}} \Delta_{A_k}$, $A_k \in \mathcal{L}$:*

(a) $\Gamma^\circ = \Delta_{A_0}$, where Γ° is the interior of Γ in $(\mathfrak{G}, \tau)$ and $A_0 = \bigcap_{k \in \mathbb{N}} A_k$.
(b) Both Γ and Γ° are closed.
(c) The boundary $\partial \Gamma = \overline{\Gamma} \setminus \Gamma^\circ$ of Γ is characterised by

$$\partial \Gamma = \left\{ \omega \in \mathfrak{G} : \omega(A_k) = 1 \text{ for all } k \in \mathbb{N} \text{ but } \omega(A_0) = 0 \right\}.$$

*(d) If in addition $A_{k+1} \subset A_k$, $k \in \mathbb{N}$, $A_k \setminus A_{k+1} \notin \mathcal{N}$ for infinitely many k, then $\partial \Gamma$
is uncountable.*

   (ii) If $\Lambda = \bigcup_{k \in \mathbb{N}} \Delta_{B_k}$, $B_k \in \mathcal{L}$:

(a) $\overline{\Lambda} = \Delta_{B^0}$, where $\overline{\Lambda}$ is the closure of Λ in $(\mathfrak{G}, \tau)$ and $B^0 = \bigcup_{k \in \mathbb{N}} B_k$.
(b) Both Λ and $\overline{\Lambda}$ are open.
(c) The boundary $\partial \Lambda = \overline{\Lambda} \setminus \Lambda^{\circ}$ of Λ is characterised by

$$\partial \Lambda = \Big\{ \omega : \omega(B_k) = 0 \text{ for all } k \in \mathbb{N} \text{ but } \omega(B^0) = 1 \Big\}.$$

*(d) If, in addition, $B_k \subset B_{k+1}$, $k \in \mathbb{N}$, and $B_{k+1} \setminus B_k \notin \mathcal{N}$ for infinitely many k, then
$\partial \Lambda$ is uncountable.*

Proof $(i)(a)$ Since Δ_{A_0} is open, it follows from (7.1b) that

$$\Delta_{A_0} = \Delta_{\bigcap_k A_k} \subset \Big(\bigcap_k \Delta_{A_k} \Big)^{\circ} = \Gamma^{\circ}.$$

Suppose $\omega \in \Gamma^{\circ} \setminus \Delta_{A_0}$ and let $E \in \mathcal{L} \setminus \mathcal{N}$ be such that $\omega \in \Delta_E \subset \Gamma$. It follows that
$E \setminus A_{k_0} \notin \mathcal{N}$ for some $k_0 \in \mathbb{N}$, for otherwise

$$E \setminus \Big(\bigcap_k A_k \Big) = \bigcup_k (E \setminus A_k) \in \mathcal{N},$$

whence

$$1 = \omega(E) = \omega(E \setminus A_0) + \omega(E \cap A_0) = \omega\Big(E \setminus \Big(\bigcap_k A_k \Big) \Big) + \omega(E \cap A_0) = 0,$$

a contradiction. Since $E \setminus A_{k_0} \notin \mathcal{N}$, it follows from Theorem 5.6 that there exists
$\omega_0 \in \mathfrak{G}$ with $\omega_0(E \setminus A_{k_0}) = 1$. This implies that $\omega_0(A_{k_0}) = 0$ and hence that $\omega_0 \notin \Gamma$.
But $\omega_0(E \setminus A_{k_0}) = 1$ also implies that $\omega_0 \in \Delta_E \subset \Gamma$. This contradiction implies that
$\Gamma^{\circ} \setminus \Delta_{\bigcap_k A_k}$ is empty and hence that $\Delta_{A_0} = \Gamma^{\circ}$.

   $(i)(b, c)$ Since Γ is an intersection of closed sets, $\Gamma = \overline{\Gamma}$ is closed, Γ° is closed
by Lemma 7.5 and the formula for the boundary $\partial \Gamma = \Gamma \setminus \Gamma^{\circ}$ follows.

   $(i)(d)$ Now suppose $A_{k+1} \subset A_k$, $k \in \mathbb{N}$, and $A_k \setminus A_{k+1} \notin \mathcal{N}$ for infinitely many
k. Then there is no loss since, by Lemma 7.1, $A_k \setminus A_{k+1} \in \mathcal{N}$ implies $\Delta_{A_k} = \Delta_{A_{k+1}}$,
in assuming that $F_k := A_k \setminus A_{k+1} \notin \mathcal{N}$ for all k. Since the F_k are non-null and
disjoint, and since $\bigcup_{j \geq k} F_j \subset A_k$, Corollary 5.7 implies the existence of a family
$\{\omega_a : a \in (0,1)\} \subset \mathfrak{G}$ of distinct finitely additive measures with $\omega_a \Big(\bigcup_{j \geq k} F_j \Big) =$
$\omega_a(A_k) = 1$, $k \in \mathbb{N}$. Hence $\omega_a \in \Gamma$ for all $a \in (0,1)$. However

$$A_0 \cap \Big(\bigcup_{j \geq k} F_j \Big) = A_0 \cap \Big(A_k \setminus A_0 \Big) = \emptyset.$$

Therefore, $\omega_a(A_0) = 0$ for all $a \in \mathbb{R}$ and hence $\omega_a \in \Gamma \setminus \Gamma_0 = \partial\Gamma$ for all $a \in (0, 1)$, as required.

(ii) For the statements about Λ, let $A_k = X \setminus B_k$ and $\Gamma = \mathfrak{G} \setminus \Lambda$ in part (i). This completes the proof. $\square$

The next result generalises the first part of Theorem 7.10 since Γ is a closed G_δ-set (a closed intersection of countably many open sets). $\square$

Corollary 7.11 *If $\mathcal{H} \subset \mathfrak{G}$ is a closed G_δ-set, there exists $\{A_k\} \subset \mathcal{L}$ with $A_{k+1} \subset A_k$ such that $\mathcal{H} = \bigcap_{k\in\mathbb{N}} \Delta_{A_k}$ and $\mathcal{H}^\circ = \Delta_{A_0}$ where $A_0 = \bigcap_k A_k$. The boundary $\partial\mathcal{H}$ is uncountable if it is non-empty, equivalently if $\mathcal{H} \notin \tau$.*

Proof Since $(\mathfrak{G}, \tau)$ is compact and Hausdorff, it is normal (see Sect. 2.10) and since $\mathcal{H}$ is a closed G_δ-set, by Lemma 2.62 there exists a non-negative continuous function $f : \mathfrak{G} \to [0, 1/2]$ with $\mathcal{H} = f^{-1}(\{0\})$. By Theorem 7.4, for some non-negative $u \in L_\infty(X, \mathcal{L}, \lambda)$,

$$f(\omega) = \int_X u \, d\omega \text{ for all } \omega \in \mathfrak{G}.$$

Let $A_k = \{x \in X : u(x) \leqslant 1/k\}$, $k \in \mathbb{N}$. Then $A_{k+1} \subset A_k$ and if $\omega \in \mathcal{H}$,

$$0 = f(\omega) = \int_X u \, d\omega = \int_{A_k} u \, d\omega + \int_{X\setminus A_k} u \, d\omega \geqslant k^{-1}\omega(X \setminus A_k) \geqslant 0,$$

and it follows that $\omega(A_k) = 1$ for all $k \in \mathbb{N}$. If $\omega \notin \mathcal{H}$ let $k \in \mathbb{N}$ be sufficiently large that $1/k < f(\omega)$. Then

$$0 < f(\omega) - \frac{1}{k} \leqslant \left(\int_{A_k} u \, d\omega - \frac{1}{k}\right) + \int_{X\setminus A_k} u \, d\omega \leqslant \|u\|_\infty \omega(X \setminus A_k)$$

and since $\omega \in \mathfrak{G}$ it follows that $\omega(A_k) = 0$ for all k sufficiently large if $\omega \notin \mathcal{H}$. Hence

$$\mathcal{H} = \left\{\omega \in \mathfrak{G} : \omega(A_k) = 1 \text{ for all } k \in \mathbb{N}\right\} = \bigcap_{k\in\mathbb{N}} \Delta_{A_k}. \qquad (7.6)$$

Since $\mathcal{H} = \bigcap_{k\in\mathbb{N}} \Delta_{A_k}$, by Theorem 7.10(i) $\mathcal{H}^\circ = \Delta_{\bigcap_{k\in\mathbb{N}} A_k}$.

Finally, note that since $\mathcal{H}$ is closed its boundary $\partial\mathcal{H} = \mathcal{H} \setminus \mathcal{H}^\circ$ is non-empty if and only if $\mathcal{H}$ is not open. If $\mathcal{H}$ is not open it follows that $A_k \setminus A_{k+1} \notin \mathcal{N}$ for infinitely many k, for otherwise, by Lemma 7.1, the collection of sets $\{\Delta_{A_k}\}$ is finite and so $\mathcal{H} = \bigcap_k \Delta_{A_k}$ is open. Thus when $\partial\mathcal{H}$ is non-empty $\{A_k\}$ satisfies the hypothesis of Theorem 7.10(i)(d) and $\partial\mathcal{H}$ is uncountable. $\square$

7.4 𝔊 and the Weak* Topology on $L_\infty^*(X, \mathcal{L}, \lambda)$

Recall from Definition 2.47 that a set $U \subset L_\infty(X, \mathcal{L}, \lambda)^*$ is open in the weak* topology if and only if for every $f_0 \in U$ there exists $u_1, \cdots, u_n \in L_\infty(X, \mathcal{L}, \lambda)$ and $\epsilon > 0$ such that

$$\bigcap_{j=1}^{n} \left\{ f \in L_\infty(X, \mathcal{L}, \lambda)^* : |(f - f_0)(u_j)| < \epsilon \right\} \subset U.$$

By analogy say $W \subset L_\infty^*(X, \mathcal{L}, \lambda)$ is weak* open in $L_\infty^*(X, \mathcal{L}, \lambda)$ if and only if the corresponding set of elements of $L_\infty(X, \mathcal{L}, \lambda)^*$, defined by Theorem 3.1, is open in the weak* topology of $L_\infty(X, \mathcal{L}, \lambda)^*$. In other words

$$\mathcal{W} := \left\{ W_{\epsilon_0, u_0, \nu_0} : \epsilon_0 > 0, u_0 \in L_\infty(X, \mathcal{L}, \lambda), \nu_0 \in L_\infty^*(X, \mathcal{L}, \lambda) \right\}, \qquad (7.7a)$$

where

$$W_{\epsilon_0, u_0, \nu_0} = \left\{ \nu \in L_\infty^*(X, \mathcal{L}, \lambda) : \left| \int_X u_0 \, d\nu - \int_X u_0 \, d\nu_0 \right| < \epsilon_0 \right\}, \qquad (7.7b)$$

is a sub-base for the weak* topology on $L_\infty^*(X, \mathcal{L}, \lambda)$.

Lemma 7.12 *The topology τ on 𝔊 coincides with the restriction to 𝔊 of the weak* topology on $L_\infty^*(X, \mathcal{L}, \lambda)$.*

Proof It suffices to show that when $\omega_0 \in \Delta_{A_0} \in \mathcal{T}$ (Definition 7.2), there exists $W_0 \in \mathcal{W}$ (see (7.7)) with $\omega_0 \in W_0 \cap 𝔊 \subset \Delta_{A_0}$, and when $\omega_0 \in W_0 \cap 𝔊$, $W_0 \in \mathcal{W}$, there exists $A_0 \in \mathcal{L}$ with $\omega_0 \in \Delta_{A_0} \subset W_0 \cap 𝔊$.

First suppose $\omega_0 \in \Delta_{A_0}$ and let

$$W_0 = \left\{ \nu \in L_\infty^*(X, \mathcal{L}, \lambda) : \left| \int_X \chi_{A_0} \, d\omega_0 - \int_X \chi_{A_0} \, d\nu \right| < \frac{1}{2} \right\} \in \mathcal{W}.$$

Then $\omega \in W_0 \cap 𝔊$ implies that $|\omega_0(A_0) - \omega(A_0)| < \frac{1}{2}$. Hence $\omega(A_0) = 1$ since $\omega_0(A_0) = 1$, and therefore $\omega_0 \in W_0 \cap 𝔊 \subset \Delta_{A_0}$.

Now let $\omega_0 \in 𝔊$ and, for any $u_0 \in L_\infty(X, \mathcal{L}, \lambda)$ and $\epsilon_0 > 0$, let

$$W_0 = \left\{ \nu \in L_\infty^*(X, \mathcal{L}, \lambda) : \left| \int_X u_0 \, d\nu - \int_X u_0 \, d\omega_0 \right| < \epsilon_0 \right\} \in \mathcal{W}.$$

Let $\alpha_0 = \int_X u_0 \, d\omega_0$ and put $A_0 = \{x \in X : |u_0(x) - \alpha_0| < \epsilon_0/2\}$. Then $\omega_0 \in \Delta_{A_0}$ by Theorem 6.2 and $\{x \in X : |u_0(x) - \hat{\alpha}| < \epsilon_0/2\} \cap A_0 \in \mathcal{N}$ if $|\hat{\alpha} - \alpha_0| \geq \epsilon_0$.

Therefore $\omega(\{x \in X : |u_0(x) - \hat{\alpha}| < \epsilon_0/2\}) = 0$, and consequently $\int_X u_0 \, d\omega \neq \hat{\alpha}$, if $\omega \in \Delta_{A_0}$ and $|\hat{\alpha} - \alpha_0| \geq \epsilon_0$. It follows that $\left| \int_X u_0 \, d\omega - \alpha_0 \right| < \epsilon_0$ if $\omega \in \Delta_{A_0}$. Hence $\omega_0 \in \Delta_{A_0} \subset W_0 \cap 𝔊$, as required. $\qquad \square$

Lemma 7.13 $\mathfrak{G}$ *is closed in the weak* topology of* $L_\infty^*(X, \mathcal{L}, \lambda)$.

Proof It suffices to show that $L_\infty^*(X, \mathcal{L}, \lambda) \setminus \mathfrak{G}$ is open. If $\nu_0 \in L_\infty^*(X, \mathcal{L}, \lambda) \setminus \mathfrak{G}$ there exists $A \in \mathcal{L}$ with $\nu_0(A) = a \notin \{0, 1\}$. Let $2\epsilon = \min\{|a|, |1 - a|\} > 0$ and suppose $\nu \in L_\infty^*(X, \mathcal{L}, \lambda)$ is in the weak* open neighbourhood of ν_0 given by

$$V = \left\{ \nu \in L_\infty^*(X, \mathcal{L}, \lambda) : \left| \int_X \chi_A \, d\nu - \int_X \chi_A \, d\nu_0 \right| < \epsilon \right\}.$$

Then (see (6.2d)) $\nu \in V$ implies that $|\nu(A) - \nu_0(A)| < \epsilon$, which means $\nu(A) \notin \{0, 1\}$, and hence $\nu \notin \mathfrak{G}$. Therefore $L_\infty^*(X, \mathcal{L}, \lambda) \setminus \mathfrak{G}$ is weak* open, and $\mathfrak{G}$ is weak* closed. $\qquad\square$

It follows from Theorem 2.48 (Alaoglu) that $(\mathfrak{G}, \tau)$ is compact. For a direct proof, see Theorem 7.3.

7.5 $\mathfrak{G}$ as Extreme Points

By Definition 2.52 f is an extreme point of B^*, the closed unit ball in $L_\infty(X, \mathcal{L}, \lambda)^*$, if and only if for $f_1, f_2 \in B^*$ and $\alpha \in (0, 1)$

$$f(u) = \alpha f_1(u) + (1 - \alpha) f_2(u) \text{ for all } u \in L_\infty(X, \mathcal{L}, \lambda)$$

implies that $f = f_1 = f_2$. Let

$$\nu, \nu_1, \nu_2 \in U^* = \{\nu \in L_\infty^*(X, \mathcal{L}, \lambda) : |\nu|(X) \leqslant 1\}$$

be defined in terms of f, f_1, $f_2 \in B^*$ by Theorem 3.1. Then a necessary and sufficient condition for f to be an extreme point of B^* is that for $\alpha \in (0, 1)$

$$\int_X u \, d\nu = \alpha \int_X u \, d\nu_1 + (1 - \alpha) \int_X u \, d\nu_2 \text{ for all } u \in L_\infty(X, \mathcal{L}, \lambda) \qquad (7.8a)$$

implies that $\nu = \nu_1 = \nu_2$. Since simple functions are dense in $L_\infty(X, \mathcal{L}, \lambda)$, it is immediate that (7.8a) holds if and only if

$$\nu(E) = \alpha\nu_1(E) + (1 - \alpha)\nu_2(E) \text{ for all } E \in \mathcal{L}. \qquad (7.8b)$$

It is obvious that f is on the boundary of B^*, and so $\|f\| = |\nu|(X) = 1$, if f is extreme. But more can be said.

Lemma 7.14 *Let* $f \in L_\infty(X, \mathcal{L}, \lambda)^*$ *and* $\nu \in L_\infty^*(X, \mathcal{L}, \lambda)$ *be related by (3.1b) in Theorem 3.1. Then* f *is an extreme point of* B^* *if and only if* ν *or* $-\nu$ *is in* $\mathfrak{G}$.

Proof First suppose f is an extreme point of B^*. Then $\|f\| = |\nu|(X) = 1$: If ν is not one-signed then $|\nu| = \nu^+ + \nu^-$ where $\nu^+ \wedge \nu^- = 0$ and $\nu^{\pm}(X) \in (0, 1)$. Let $0 < \epsilon_0 = \frac{1}{2}\min\{\nu^+(X), \nu^-(X)\}$ and, by (4.1c), choose $A \in \mathcal{L}$ such that $\nu^+(X \setminus A) + \nu^-(A) = \epsilon < \epsilon_0$. If $\nu(A) = 0$ then $\nu^+(X) = \nu^+(X \setminus A) + \nu^+(A) = \nu^+(X \setminus A) + \nu^-(A) = \epsilon < \epsilon_0$, which is false. So $\nu(A) \neq 0$ and hence $|\nu|(A) > 0$. If $|\nu|(A) = 1$ then

$$\nu^-(X) = 1 - \nu^+(X) = \nu^-(A) + \nu^+(A) - \nu^+(X)$$
$$= \nu^-(A) - \nu^+(X \setminus A) \leqslant \nu^-(A) + \nu^+(X \setminus A) = \epsilon < \epsilon_0$$

which is false. So $|\nu|(A) \in (0, 1)$. Let

$$\nu_1(E) = \frac{\nu(A \cap E)}{|\nu|(A)}, \quad \nu_2(E) = \frac{\nu((X \setminus A) \cap E)}{|\nu|(X \setminus A)} \quad \text{for all } E \in \mathcal{L}.$$

Then $\nu_1, \nu_2 \in U^*$ and, for all $E \in \mathcal{L}$,

$$\nu(E) = \alpha\nu_1(E) + (1 - \alpha)\nu_2(E), \text{ where } \alpha = |\nu|(A), \ (1 - \alpha) = |\nu|(X \setminus A).$$

Since $\alpha \in (0, 1)$, $\nu_1(A) = \nu(A)/|\nu|(A) \neq 0$ and $\nu_2(A) = 0$, this shows that f is not an extreme element of B^* if ν is not one-signed.

So suppose ν is one-signed, say $0 \leqslant \nu \in U^*$ (for $\nu \leqslant 0$ replace ν with $-\nu$), but $\nu \notin \mathfrak{G}$. Then there exists $A \in \mathcal{L}$ with $\nu(A) \in (0, 1)$. Let

$$\nu_1(E) = \frac{\nu(A \cap E)}{\nu(A)}, \quad \nu_2(E) = \frac{\nu((X \setminus A) \cap E)}{\nu(X \setminus A)} \quad \text{for all } E \in \mathcal{L}.$$

Then $\nu_1, \nu_2 \in U^*$, $\nu(E) = \alpha\nu_1(E) + (1 - \alpha)\nu_2(E)$ for all $E \in \mathcal{L}$, where $\alpha = \nu(A)$, $(1 - \alpha) = \nu(X \setminus A)$. Since $\nu_1(A) = 1 \neq 0 = \nu_2(A)$, ν is not extreme in U^*. Hence $\pm\nu \in \mathfrak{G}$ if f is an extreme point of B^*.

Now to show that f is an extreme point of B^* if $\nu \in \mathfrak{G}$, suppose that $\nu_1, \nu_2 \in U^*$ and for all $E \in \mathcal{L}$,

$$\nu(E) = \alpha\nu_1(E) + (1 - \alpha)\nu_2(E), \quad \alpha \in (0, 1), \quad \nu_1, \nu_2 \in U^*.$$

Then $\nu \geqslant 0$ and if $\nu(E) = 1$,

$$1 = \nu(E) = \alpha\nu_1(E) + (1 - \alpha)\nu_2(E) \leqslant \alpha|\nu_1|(X) + (1 - \alpha)|\nu_2|(X) \leqslant 1$$

which implies that $\nu_1(E) = \nu_2(E) = \nu(E) = 1$. In particular $\nu_1(X) = \nu_2(X) = 1$. If $\nu(E) = 0$ then $\nu(X \setminus E) = 1$ and so $\nu_1(X \setminus E) = \nu_2(X \setminus E) = 1$, whence $\nu_1(E) = \nu_2(E) = \nu(E) = 0$. Thus $\nu = \nu_1 = \nu_2$ which shows that f is an extreme point of B^* if $\nu \in \mathfrak{G}$. $\qquad\square$

Remark 7.15 Because of Theorem 2.55 (Rainwater), Lemma 7.14 implies that $\mathfrak{G}$ has property (W) in the Introduction. In Sect. 8, this conclusion is derived independently, from the isometric isomorphism in Theorem 7.4.   □

Let (X, ϱ) be a Hausdorff space, $\mathcal{B}$ its Borel σ-algebra and let $\mathfrak{D}$ denote the set of regular Borel measures (Definition 2.12) δ that take only values 0 and 1 on $\mathcal{B}$ with $\delta(X) = 1$.

Lemma 7.16 *For $\delta \in \mathfrak{D}$ there exists a unique $x_0 \in X$ such that $\delta = \delta_{x_0}$, the Dirac measure introduced in Sect. 2.4.*

Proof Since $\delta \in \mathfrak{D}$ is regular, there is a compact $K \subset X$ with $\delta(K) = 1$. Let K_0 denote the intersection of all such compact K and note that $K_0 \neq \emptyset$ because these compact sets have the finite intersection property. If $x_0, y_0 \in K_0$ are distinct, then at least one of $\{x_0\}, \{y_0\}$ has zero measure. If $\delta(\{y_0\}) = 0$, by the regularity of δ there is an open set G_{y_0} with $y_0 \in G_{y_0}$ and $\delta(G_{y_0}) = 0$. So $\delta(K \setminus G_{y_0}) = 1$ for all compact K with $\delta(K) = 1$. This contradicts the fact that $y_0 \in K_0$. Hence K_0 is a singleton, $\{x_0\}$ say. If $\delta(\{x_0\}) = 0$, the same argument implies that $x_0 \notin K_0$. So $K_0 = \{x_0\}$ and $\delta = \delta_{x_0}$.   □

Chapter 8
Weak Convergence in $L_\infty(X, \mathcal{L}, \lambda)$

In this chapter, theory so far developed yields a necessary and sufficient condition in terms of behaviour λ-almost everywhere for a bounded sequence that is pointwise convergent to be weakly convergent in $L_\infty(X, \mathcal{L}, \lambda)$. The resulting test is illustrated by examples.

8.1 Weakly Convergent Sequences

Since, by Sect. 7.2, $L_\infty(X, \mathcal{L}, \lambda)$ and $C(\mathfrak{G}, \tau)$ are isometrically isomorphic, their duals are isometrically isomorphic and $u_k \rightharpoonup u_0$ in $L_\infty(X, \mathcal{L}, \lambda)$ if and only if $L[u_k] \rightharpoonup L[u_0]$ in $C(\mathfrak{G}, \tau)$. Since $(\mathfrak{G}, \tau)$ is a compact Hausdorff space, by (V) in the Introduction $L[u_k] \rightharpoonup L[u_0]$ in $C(\mathfrak{G}, \tau)$ if and only if $\{\|L[u_k]\|_{C(\mathfrak{G}, \tau)}\}$ is bounded and $L[u_k] \to L[u_0]$ pointwise on $\mathfrak{G}$. In other words, $u_k \rightharpoonup u_0$ in $L_\infty(X, \mathcal{L}, \lambda)$ if and only if for some M

$$\|u_k\|_\infty \leqslant M \text{ and } \int_X u_k \, d\omega \to \int_X u_0 \, d\omega \text{ as } k \to \infty \text{ for all } \omega \in \mathfrak{G}. \qquad (8.1)$$

Thus the collection $\mathfrak{G}$ of finitely additive measures plays a rôle for weak convergence in $L_\infty(X, \mathcal{L}, \lambda)$ analogous to that of σ-additive Dirac measures for weak convergence in spaces of continuous functions. The sequential weak continuity of composition operators is an immediate consequence, in contrast to what happens for weak convergence in $L_p(X, \mathcal{L}, \lambda)$, $1 \leqslant p < \infty$. (When $u_k(x) = \sin(kx)$ in $L_p(0, 2\pi)$, $u_k \rightharpoonup 0$ but $|u_k| \nrightarrow 0$.)

Theorem 8.1 *If $u_k \rightharpoonup u_0$ in $L_\infty(X, \mathcal{L}, \lambda)$ as $k \to \infty$ and $F : \mathbb{R} \to \mathbb{R}$ is continuous, then $F(u_k) \rightharpoonup F(u_0)$ in $L_\infty(X, \mathcal{L}, \lambda)$.*

Proof This is immediate from Theorem 6.2(c) and (8.1). $\qquad \square$

J. Toland, *The Dual of $L_\infty(X, \mathcal{L}, \lambda)$, Finitely Additive Measures and Weak Convergence*, SpringerBriefs in Mathematics, https://doi.org/10.1007/978-3-030-34732-1_8

Corollary 8.2 *In $L_\infty(X, \mathcal{L}, \lambda)$, $u_k \rightharpoonup 0$ if and only if $|u_k| \rightharpoonup 0$.*

Proof "Only if" follows from Theorem 8.1 and "if" from (6.10a) and (8.1). $\square$

Recall from Remark 2.9 that if A is an atom, $\lambda(A) \in (0, \infty)$ and $u \in L_\infty(X, \mathcal{L}, \lambda)$ is constant on A. Recall also from Theorem 5.9 that $\omega \in \mathfrak{G}$ is either purely finitely additive or, for all $E \in \mathcal{L}$, $\omega(E)$ is a constant multiple of $\lambda(E_\omega \cap E)$ where E_ω is an atom. Hence from (8.1) if $\{u_k\}$ is bounded in $L_\infty(X, \mathcal{L}, \lambda)$, $u_k \rightharpoonup u$ in $L_\infty(X, \mathcal{L}, \lambda)$ if and only if

$$u_k(A) \to u(A) \text{ for all atoms } A \in \mathcal{L}, \tag{8.2a}$$

and

$$\int_X u_k \, d\mu \to \int_X u \, d\mu \text{ for all purely finitely additive } \mu \in \mathfrak{G}. \tag{8.2b}$$

Of course many familiar measure spaces have no atoms (Example 2.15), but when they do (8.2a) is equivalent to

$$\int_X \chi_A u_k \, d\lambda \to \int_X \chi_A u \, d\lambda \text{ for all atoms } A.$$

Hence for bounded sequences (8.2a) holds if $u_k \overset{*}{\rightharpoonup} u$, and in particular if $u_k(x) \to u(x)$ for λ-almost all $x \in X$.

Since $(X, \mathcal{L}, \lambda)$ is σ-finite, $L_\infty(X, \mathcal{L}, \lambda)$ is the dual of $L_1(X, \mathcal{L}, \lambda)$ [8, Thm. 4.4.1] and hence by Theorem 2.48 (Alaoglu) the closed unit ball in $L_\infty(X, \mathcal{L}, \lambda)$ is compact in the weak* topology of $L_1(X, \mathcal{L}, \lambda)^*$ (Definition 2.47). If in addition $L_1(X, \mathcal{L}, \lambda)$ is a separable metric space, by Remark 2.51 the closed unit ball in $L_\infty(X, \mathcal{L}, \lambda)$ with the weak* topology is metrisable, and hence weak* sequentially compact (Definition 2.49). Thus a bounded sequence in $L_\infty(X, \mathcal{L}, \lambda)$ has a weak* convergent subsequence which therefore satisfies (8.2a). (For example $(X, \mathcal{L}, \lambda)$ is σ-finite and $L_1(X, \mathcal{L}, \lambda)$ is separable when $X \subset \mathbb{R}^n$ and λ is any Borel measure on $\mathbb{R}^n$ which is finite on bounded sets [8, Cor. 4.2.2].)

In circumstances such as these, deciding whether a bounded sequence is weakly convergent or has a weakly convergent subsequence is mainly a question of deciding whether (8.2b) is satisfied. Theorem 8.7 settles this issue by characterising sequences that converge weakly to 0 in terms of their λ-almost everywhere pointwise behaviour. To set the scene, recall some necessary conditions for the weak convergence of sequences in $L_\infty(X, \mathcal{L}, \lambda)$.

8.2 Pointwise Characterisation

Lemma 8.3 *When $u_k \rightharpoonup u_0$ in $L_\infty(X, \mathcal{L}, \lambda)$, there is a subsequence with $u_{k_j}(x) \to u_0(x)$ for λ-almost all $x \in X$.*

Proof Since $(X, \mathcal{L}, \lambda)$ is σ-finite there exists $f \in L_1(X, \mathcal{L}, \lambda)$ which is positive λ-almost everywhere and, by Corollary 8.2, $|u_k - u_0|f \to 0$ in $L_1(X, \mathcal{L}, \lambda)$. Hence a subsequence $u_{k_j}(x) \to u_0(x)$ for λ-almost all x. $\square$

In a metric space, let $B(x, r)$ denote the ball with centre x and radius r.

Lemma 8.4 *Suppose (X, ρ) is a metric space and for $u \in L_\infty(X, \mathcal{B}, \lambda)$, there exists a set $E(u) \in \mathcal{L}$ such that $\lambda(X \setminus E(u)) = 0$,*

$$\underline{u}(x) := \lim_{0<r\to0} \frac{1}{\lambda(B(x,r))} \int_{B(x,r)} u \, d\lambda \text{ exists for all } x \in E(u), \qquad (8.3)$$

and $\underline{u}(x) = u(x)$, $x \in E(u)$. Then $u_k \rightharpoonup u_0$ in $L_\infty(X, \mathcal{B}, \lambda)$ implies that $u_k(x) \to u_0(x)$ for λ-almost all $x \in X$.

Remark 8.5 Hypothesis (8.3) means that λ-almost all $x \in X$ is a Lebesgue point of u [28, 7.6 & Th. 7.7]. According to Heinonen [18, Thm. 1.8 & Remark 1.13] this is not very restrictive. For example, it holds when λ is a regular Borel measure on $\mathcal{B}$ which is doubling on (X, ρ) (i.e. balls have finite positive measure and there exists a constant C such that $\lambda(B(x, 2r)) \leqslant C\lambda(B(x, r))$, or on $\mathbb{R}^n$ with the standard metric when λ is a Radon measure, finite on compact sets and positive on balls. $\square$

Proof Let $E = \bigcap_0^\infty E(u_k)$, which has full measure, and let V denote the linear span in $L_\infty(X, \mathcal{B}, \lambda)$ of $\{u_k : k \in \mathbb{N}_0\}$. Now for fixed $x \in E$ define a linear functional ℓ_x on V by

$$\ell_x(v) = \underline{v}(x), \quad v \in V.$$

Since $|\ell_x(v)| \leqslant \|v\|_\infty$, by the Hahn–Banach theorem there exists $L_x \in L_\infty(X, \mathcal{B}, \lambda)^*$ with $L_x(v) = \ell_x(v)$ for all $v \in V$. Therefore $u_k \rightharpoonup u_0$ implies that $u_k(x) = \ell_x(u_k) = L_x(u_k) \to L_x(u_0) = u_0(x)$ for all $x \in E$, as required. $\square$

Example 8.6 To see that the converse is false let $X = (0, 1)$ and, for $k \in \mathbb{N}$, define u_k on $(0, 1)$ by $u_k(x) = 0$, $x \in [1/2k, 1)$; $u_k(x) = 1$, $x \in (0, 1/4k]$; u_k linear on $[1/4k, 1/2k]$. Then $\{\|u_k\|_\infty\}$ is bounded, u_k is continuous on X and $u_k(x) \to 0$ monotonically for all $x \in X$ as $k \to \infty$. But by Theorem 5.6 there exists $\omega \in \mathfrak{G}$ with $\omega(E_\ell) = 1$, where $E_\ell = (0, 1/4\ell)$ for each ℓ. Therefore, by Theorem 6.2, $\int_X u_k \, d\omega = 1$ for all k and hence $u_k \not\rightharpoonup 0$ in $L_\infty(X, \mathcal{L}, \lambda)$. Hence u_k is not weakly convergent to 0 (indeed not weakly convergent to any limit) in $L_\infty(X, \mathcal{L}, \lambda)$. $\square$

If $\{k_j\} \subset \mathbb{N}$ is strictly increasing and $u_k \rightharpoonup 0$ in $L_\infty(X, \mathcal{L}, \lambda)$, by Corollary 2.46 of Mazur's theorem and Corollary 8.2 there exists $\{\overline{u}_i\}$ in the convex hull of $\{|u_{k_j}| : j \in \mathbb{N}\}$ with

$$\|\overline{u}_i\|_\infty \to 0 \text{ as } i \to \infty, \quad \overline{u}_i = \sum_{j=1}^{m_i} \gamma_j^i |u_{k_j}|, \quad \gamma_j^i \in [0, 1] \text{ and } \sum_{j=1}^{m_i} \gamma_j^i = 1, \ m_i \in \mathbb{N}.$$

Since γ_j^i may be zero, there is no loss in assuming that $\{m_i\}$ is increasing. Therefore, for any strictly increasing sequence $\{k_j\}$,

$$0 \leqslant w_i(x) := \min\left\{|u_{k_j}(x)| : j \in \{1, \cdots, m_i\}\right\} \leqslant \bar{u}_i(x), \ x \in X,$$

defines a non-increasing sequence of non-negative functions in $L_\infty(X, \mathcal{L}, \lambda)$ with $\|w_i\|_\infty \to 0$. It follows that if $u_k \rightharpoonup 0$ and $\{k_j\} \subset \mathbb{N}$ is any strictly increasing sequence, the corresponding sequence defined on X by

$$0 \leqslant v_J(x) := \min\left\{|u_{k_j}(x)| : j \in \{1, \cdots, J\}\right\} \tag{8.4}$$

is pointwise non-increasing in $L_\infty(X, \mathcal{L}, \lambda)$ and $\|v_J\|_\infty \to 0$ as $J \to \infty$. To show that this condition is sufficient as well as necessary for a sequence to be weakly convergent to 0 in $L_\infty(X, \mathcal{L}, \lambda)$, let

$$A_\alpha(u) = \left\{x \in X : |u(x)| > \alpha\right\}, \ u \in L_\infty(X, \mathcal{L}, \lambda), \ \alpha > 0.$$

The proof is independent of Mazur's theorem.

Theorem 8.7 *In $L_\infty(X, \mathcal{L}, \lambda)$ bounded sequence $\{u_k\}$ converges weakly to zero if and only if for every $\alpha > 0$ and every strictly increasing sequence $\{k_j\}$ in $\mathbb{N}$ there exists $J \in \mathbb{N}$ with*

$$\lambda\left(\bigcap_{j=1}^J A_\alpha(u_{k_j})\right) = 0. \tag{8.5}$$

This criterion is equivalent to saying that for all strictly increasing sequences $\{k_j\}$, the sequence $\{v_J\}$ in (8.4) converges strongly to zero in $L_\infty(X, \mathcal{L}, \lambda)$.

Remark 8.8 In the Borel setting of the next chapter, this criterion for weak convergence in $L_\infty(X, \mathcal{B}, \lambda)$ will be localised to arbitrary open neighbourhoods of points in the one-point compactification of X. □

Proof Suppose, for a strictly increasing sequence $\{k_j\}$ and $\alpha > 0$, that (8.5) is false for all $J \in \mathbb{N}$. Then $\mathcal{E} = \{A_\alpha(u_{k_j}) : j \in \mathbb{N}\}$ satisfies the hypothesis of Theorem 5.6 and hence there exists $\omega \in \mathfrak{G}$ such that $\omega(A_\alpha(u_{k_j})) = 1$ for all j. It follows that

$$\int_X |u_{k_j}| \, d\omega \geqslant \int_{A_\alpha(u_{k_j})} |u_{k_j}| \, d\omega \geqslant \alpha > 0 \text{ for all } j.$$

Hence $|u_k| \not\rightharpoonup 0$ by (8.1) and so, by Corollary 8.2, $u_k \not\rightharpoonup 0$.

Conversely, suppose $u_k \not\rightharpoonup 0$. Then by (8.1) and Corollary 8.2 there exists $\alpha > 0$, a strictly increasing sequence $\{k_j\} \subset \mathbb{N}$ and $\omega \in \mathfrak{G}$ such that

$$\int_X |u_{k_j}| \, d\omega =: \alpha_j > \alpha > 0 \text{ for all } j \in \mathbb{N}. \tag{8.6}$$

Now $\alpha_j - \alpha > 0$ for all j and

$$\left\{ x : \left| |u_{k_j}(x)| - \alpha_j \right| < \alpha_j - \alpha \right\} \subset \left\{ x : |u_{k_j}(x)| > \alpha \right\} = A_\alpha(u_{k_j}).$$

From (8.6) and Theorem 6.2 it follows that

$$1 = \omega\left(\left\{ x : \left| |u_{k_j}(x)| - \alpha_j \right| < \alpha_j - \alpha \right\} \right) \leqslant \omega\left(A_\alpha(u_{k_j}) \right),$$

and so $\omega(A_\alpha(u_{k_j})) = 1$ for all j. Hence $\omega(\bigcap_{j=1}^{J} A_\alpha(u_{k_j})) = 1$, for all J, since $\omega \in \mathfrak{G}$. Since ω is zero on $\mathcal{N}$ because $\omega \in \mathfrak{G}$, it follows that (8.5) is false for all J. Finally, note that for a strictly increasing sequence $\{k_j\}$ and $\alpha > 0$,

$$\lambda\left(\left\{ x : v_J(x) > \alpha \right\} \right) = \lambda\left(\left\{ x : |u_{k_j}(x)| > \alpha \text{ for all } j \in \{1, \cdots, J\} \right\} \right)$$

$$= \lambda\left(\bigcap_{j=1}^{J} A_\alpha(u_{k_j}) \right).$$

Since $v_J(x) \geqslant v_{J+1}(x) \geqslant 0$ it follows that $v_J \to 0$ in $L_\infty(X, \mathcal{L}, \lambda)$ if and only if, for every $\alpha > 0$, (8.5) holds for some J. This completes the proof. □

Dini's theorem [27, Thm. 7.13] says that on compact spaces monotone pointwise convergence of a sequence of continuous functions to a continuous function is uniform; equivalently, by (V) in the Introduction, weak and strong convergence coincide for bounded monotone sequences of continuous functions on compact spaces. An analogue in $L_\infty(X, \mathcal{L}, \lambda)$ now follows either from Dini's theorem in the light of Theorem 7.4, or directly from Theorem 8.7. The sequence in Example 8.6 converges, pointwise everywhere and monotonically, to zero, but not weakly in $L_\infty(X, \mathcal{L}, \lambda)$.

Corollary 8.9 *Suppose $\{u_k\}$ is bounded in $L_\infty(X, \mathcal{L}, \lambda)$ and $|u_k(x)| \geqslant |u_{k+1}(x)|$, $k \in \mathbb{N}$, for λ-almost all $x \in X$. Then $u_k \rightharpoonup 0$ if and only if $u_k \to 0$ in $L_\infty(X, \mathcal{L}, \lambda)$.*

Proof Suppose $u_k \rightharpoonup 0$ in $L_\infty(X, \mathcal{L}, \lambda)$. Since the monotonicity in k of $\{|u_k(x)|\}$ implies that $v_J = |u_J|$, by Theorem 8.7, $u_J \to 0$ in $L_\infty(X, \mathcal{L}, \lambda)$ as $J \to \infty$. The converse is obvious. □

Remark 8.10 Let $u_k = \chi_{A_k}$ where $\{A_k\} \subset \mathcal{L}$ and $\lambda(A_i \cap A_j) = 0$, $i \neq j$. Then, defined in terms of any increasing sequence $\{k_j\}$, $v_J = 0$ for all $J \geqslant 2$. Hence $u_k \rightharpoonup 0$ as $k \to \infty$ in $L_\infty(X, \mathcal{L}, \lambda)$ by Theorem 8.7. This is a new perspective on Corollary 3.2. □

With $\ell_\infty(\mathbb{N})$ as in Example 5.8 and Sect. 6.5, for every $j \in \mathbb{N}$ a bounded linear functional δ_j is defined on $\ell_\infty(\mathbb{N})$ by $\delta_j(u) = u(j)$, $u \in \ell_\infty(\mathbb{N})$. Hence $u_k \rightharpoonup 0$ in $\ell_\infty(\mathbb{N})$ implies that $u_k(j) \to 0$ as $k \to \infty$ for each $j \in \mathbb{N}$. However, pointwise convergence does not imply weak convergence.

Corollary 8.11 *Let $\{u_k\}$ be a sequence in $\ell_\infty(\mathbb{N})$ with $u_k(i) \to 0$ as $k \to \infty$ for each $i \in \mathbb{N}$. Then $u_k \not\to 0$ if and only if there exist $\alpha > 0$ and a strictly increasing sequence $\{k_j\}$ for which there exist strictly increasing sequences $\{i_n\}$, $\{J_n\}$ such that for all $n \in \mathbb{N}$*

$$|u_{k_j}(i_n)| > \alpha \text{ for all } j \in \{1, 2 \cdots\cdots, J_n\}. \tag{8.7}$$

Proof Suppose $u_k \not\to 0$. Then by Theorem 8.7 there exists $\alpha > 0$ and a strictly increasing sequence $\{k_j\}$ such that for all $J \in \mathbb{N}$

$$\|v_J\|_\infty > \alpha \text{ where } v_J(i) = \min\left\{|u_{k_j}(i)| : j \in \{1, \cdots, J\}\right\} \text{ for } i \in \mathbb{N}. \tag{8.8}$$

Let $J_1 = 1$ and let i_1 denote the smallest i such that $|u_{k_1}(i)| > \alpha$. Since u_{k_j} converges pointwise to 0 as $j \to \infty$, there exists a smallest $J_2 > J_1$ such that $|u_{k_j}(i_1)| \leqslant \alpha$ for all $j \geqslant J_2$. Therefore by (8.8) there exists a smallest $i > i_1$, denoted by i_2, such that

$$\min\left\{|u_{k_j}(i_2)| : j \in \{1, \cdots, J_2\}\right\} > \alpha.$$

Again there exists a smallest $J_3 > J_2$ such that $|u_{k_j}(i_2)| \leqslant \alpha$ for all $j \geqslant J_3$. Let i_3 denote the smallest $i > i_2$ such that

$$\min\left\{|u_{k_j}(i_3)| : j \in \{1, \cdots, J_3\}\right\} > \alpha.$$

By induction, for the strictly increasing sequence $\{k_j\}$ satisfying (8.8), this process yields strictly increasing sequences $\{i_n\}$ and $\{J_n\}$ with the required properties.

Conversely, if (8.7) holds it is immediate that $\|v_J\|_\infty \not\to 0$ as $J \to \infty$ and hence $u_k \not\to 0$ as $k \to \infty$ by Theorem 8.7. This completes the proof. □

Example 8.12 A sequence $\{u_j\}$ in $\ell_\infty(\mathbb{N})$ with

$$u_j(i) = \begin{cases} u_{ij} \to 0 \text{ as } j \to \infty, & i < j \\ \geqslant \alpha > 0, & i \geqslant j \end{cases}, \quad i \in \mathbb{N}.$$

is pointwise convergent to 0, but not weakly convergent in $\ell_\infty(\mathbb{N})$ by Corollary 8.11 with $i_n = J_n = n$. □

8.3 Applications of Theorem 8.7

Example 8.13 In Example 2.29 $u_k \to 0$ strongly in $L_p(0, 1)$ for all finite p, and therefore $\{u_k\}$ has a subsequence (see Sect. 2.27) which converges almost everywhere to 0, yet $\limsup_{k\to\infty} u_k(x) = 1$ for all $x \in (0, 1)$. To see that $u_k \not\to 0$ in $L_\infty(0, 1)$ let

$$k_j = 1 + \frac{j(j+1)}{2} \text{ and put } I_{k_j} = \left(0, \frac{1}{j+1}\right).$$

Then

$$u_{k_j}(x) \geqslant 1, \ x \in \left(0, \frac{1}{J+1}\right) \text{ when } j \in \{1, 2, \cdots, J\},$$

and the conclusion follows from Theorem 8.7. However

$$u_{k_j} = \chi_{I_{k_j}} \text{ where } I_{k_j} = \left(\frac{2^j - 1}{2^j}, \frac{2^{j+1} - 1}{2^{j+1}}\right),$$

defines a subsequence for which $u_{k_j} \rightharpoonup 0$ by Remark 8.10, since $\{I_{k_j}\}$ is a sequence of disjoint intervals. $\qquad\square$

Example 8.14 (Step Functions) In this example $X = (-1, 1)$ with Lebesgue measure and

$$u_k = \chi_{A_k} \text{ where } A_k = \left[\frac{1}{2^{k+1}}, \frac{1}{2^k}\right).$$

Let

$$u_k^+(x) = u_k\left(x + \frac{1}{2^{(k+1)}}\right) = \chi_{A_k^+}, \text{ where } A_k^+ = \left[0, \frac{1}{2^{(k+1)}}\right),$$
$$u_k^-(x) = u_k\left(x - \frac{1}{2^{(k+1)}}\right) = \chi_{A_k^-}, \text{ where } A_k^- = \left[\frac{1}{2^k}, \frac{3}{2^{(k+1)}}\right).$$

So $\|u_k\| = \|u_k^\pm\| = 1$, and $u_k^- \rightharpoonup 0$ and $u_k \rightharpoonup 0$ by Remark 8.10 since $\{A_k\}$ and $\{A_k^-\}$ are sequences of disjoint measurable sets.

However v_J, defined in (8.4) by $\{u_k^+\}$, is 1 on $(0, 2^{-(J+1)})$ for all $J \in \mathbb{N}$, and hence $u_k^+ \not\rightharpoonup 0$ in $L_\infty(X, \mathcal{L}, \lambda)$ by Theorem 8.7. The key observation is that $\chi_{A_k} \rightharpoonup 0$ for any family $\{A_k\}$ of disjoint sets in $\mathcal{L}$. For another illustration, see Example 9.9. $\qquad\square$

Example 8.15 (Simple Functions) In $L_\infty(X, \mathcal{L}, \lambda)$ let $u_k \doteq \sum_{i=1}^\infty \alpha_i \chi_{A_k^i}$, where $\sum_{i=1}^\infty |\alpha_i| < \infty$ and, for each $i \in \mathbb{N}$, $\{A_k^i\}_{k \in \mathbb{N}}$ is a family of mutually disjoint measurable sets. Then $u_k \rightharpoonup 0$.

To see this, note that for each $x \in X$ and $i \in \mathbb{N}$ there exists at most one $k \in \mathbb{N}$, denoted, if it exists, by $\kappa(x, i)$, such that $x \in A_k^i$ if and only if $k = \kappa(x, i)$. Note also that for $\epsilon > 0$ there exists $I_\epsilon \in \mathbb{N}$ such that $\sum_{I_\epsilon+1}^\infty |\alpha_i| < \epsilon$. Hence, for any given $k \in \mathbb{N}$ and $x \in X$,

$$|u_k(x)| \leqslant \sum_{i=1}^{I_\epsilon} |\alpha_i| \chi_{A_k^i}(x) + \epsilon = \sum_{\substack{i \in \{1, \cdots, I_\epsilon\} \\ \kappa(x,i)=k}} |\alpha_i| + \epsilon.$$

Since $\{\kappa(x, i) : i \in \{1, \cdots, I_\epsilon\}\}$ has at most I_ϵ elements, there exists $k \in \{1, \cdots, I_\epsilon + 1\}$ such that $k \neq \kappa(x, i)$ for any $i \in \{1, \cdots, I_\epsilon\}$. Consequently, $\inf\{|u_k(x)| : 1 \leqslant k \leqslant I_\epsilon + 1\} \leqslant \epsilon$, independent of $x \in X$. Since this argument can be repeated with $k \in \mathbb{N}$ replaced by any strictly increasing subsequence $\{k_j\}$, $\{v_J\}$ defined in terms of that subsequence by (8.4) has $\|v_J\|_\infty \to 0$ in $L_\infty(X, \mathcal{L}, \lambda)$, and $u_k \rightharpoonup 0$ follows. $\qquad\square$

Example 8.16 (Translations) Let $u : \mathbb{R} \to \mathbb{R}$ be essentially bounded and measurable with $|u(x)| \to 0$ as $|x| \to \infty$ and let $u_k(x) = u(x+k)$. Then $u_k \rightharpoonup 0$ in $L_\infty(\mathbb{R}, \mathcal{L}, \lambda)$ where λ is Lebesgue measure on $\mathbb{R}$. To see this, for $\epsilon > 0$ suppose that $|u(x)| < \epsilon$ if $|x| > K_\epsilon$. Then for any strictly increasing $\{k_j\} \subset \mathbb{N}$, $\|v_J\|_\infty < \epsilon$ for all $J \geq 2K_\epsilon$ where $\{v_J\}$ is defined in terms of $\{u_{k_j}\}$ by (8.4), and the result follows.

For $u : \mathbb{R} \to \mathbb{R}$ essentially bounded and measurable with $u(x) \to 0$ as $x \to \infty$ and $u(x) \to 1$ as $x \to -\infty$, let $u_k(x) = u(x+k)$. Then $u_k(x) \to 0$ as $k \to \infty$ for all $x \in \mathbb{R}$, but u_k is not weakly convergent to 0 because of Theorem 8.7. However, in the notation of Definition 9.6, $u_k \rightharpoonup 0$ at every point of $\mathbb{R}$ but not, as will be seen in Chap. 9, at the point at infinity in its one-point compactification. □

Example 8.17 (Oscillatory Functions) With $X = (0, 2\pi)$ and Lebesgue measure, let $u_k(x) = \sin(1/(kx))$. Clearly $|u_k(x)| \to 0$ as $k \to \infty$ uniformly on $(\epsilon, 2\pi)$ for any $\epsilon \in (0, 2\pi)$. Therefore, if a subsequence $\{u_{k_j}\}$ is weakly convergent, its weak limit is zero.

To see that no subsequence of $\{u_k\}$ is weakly convergent to 0, consider first a strictly increasing sequence $\{k_j\}$ of natural numbers for which there exists a prime power p^m which does not divide k_j for all j. Then, for $J \in \mathbb{N}$ sufficiently large, let

$$x_J = \left\{ \frac{\pi}{p^m} \mathrm{lcm}\,\{k_1, \cdots, k_J\} \right\}^{-1} \in (0, 2\pi),$$

where lcm denotes the least common multiple. Then, for $j \in \{1, \cdots, J\}$, since $p^m \nmid k_j$ and p is prime,

$$\frac{1}{k_j x_J} = \frac{\mathrm{lcm}\,\{k_1, \cdots, k_J\}}{p^m k_j} \pi \text{ where}$$
$$\frac{\mathrm{lcm}\,\{k_1, \cdots, k_J\}}{k_j} = r \bmod p^m, \quad r \in \{1, \cdots, p^m - 1\}.$$

It follows that $|u_{k_j}(x_J)| \geq |\sin(\pi/p^m)| > 0$, independent of $j \in \{1, \cdots, J\}$ and, since u_{k_j} is continuous at x_J, that $\|v_J\|_{L_\infty(X, \mathcal{L}, \lambda)} \geq |\sin(\pi/p^m)| > 0$ for all J sufficiently large. By Theorem 8.7, this shows that $u_{k_j} \not\rightharpoonup 0$ if $\{k_j\}$ has a subsequence $\{k_j'\}$ for which $p^m \nmid k_j'$ for all $j \in \mathbb{N}$.

Note that if $\{k_j\}$ has no such subsequence for any prime p and $m \in \mathbb{N}$, then every $K \in \mathbb{N}$ is a divisor of k_j for all j sufficiently large, how large depending on K. Consequently, if $u_{k_j} \rightharpoonup 0$, $\{k_j\}$ has subsequence $\{k_j'\}$ such that

$$2k_j' \text{ divides } k_{j+1}' \text{ and } k_{j+1}' \geq 2 \sum_{i=1}^{j} k_i' \text{ for all } j \in \mathbb{N}.$$

Now for $J \in \mathbb{N}$ let

$$x_J = \frac{2}{\pi} \left(\sum_{i=1}^{J} k_i' \right)^{-1}.$$

Then the properties of $\{k'_j\}$ imply that for $j \in \{1, 2, \cdots, J\}$,

$$\frac{1}{k'_j x_J} = \frac{\left(\sum_{i=1}^{J} k'_i\right)}{k'_j}\left(\frac{\pi}{2}\right) = \frac{\left(\sum_{i=j+1}^{J} k'_i\right) + k'_j + \left(\sum_{i=1}^{j-1} k'_i\right)}{k'_j}\left(\frac{\pi}{2}\right)$$

$$= \left(2N + 1 + z\right)\left(\frac{\pi}{2}\right) \text{ where } N \in \mathbb{N} \text{ and } |z| < \frac{1}{2}.$$

Hence

$$|u_{k'_j}(x_J)| \in \left[\sin\left(\frac{\pi}{4}\right), \sin\left(\frac{\pi}{2}\right)\right] = \left[\frac{1}{\sqrt{2}}, 1\right], \ j \in \{1, \cdots, J\}.$$

It follows from Theorem 8.7 that $u_{k_j} \not\to 0$ as $j \to \infty$ since it has a subsequence $\{u_{k'_j}\}$ which generates a sequence $\{v_J\}$ with $\|v_J\|_\infty \geqslant 1/\sqrt{2}$ for all J. □

Chapter 9
L_∞^* When X is a Topological Space

This chapter considers what more can be said when (X, ϱ) is a locally compact Hausdorff space, with $\mathcal{B}$ the corresponding Borel σ-algebra and λ a measure on $\mathcal{B}$ as described in Chap. 3. In addition, here λ is assumed regular, finite on compact sets and positive on open sets. Recall from Lemma 7.16 that a regular Borel measure that takes only values 0 or 1 is a Dirac measure concentrated at a point $x_0 \in X$. Now it will be shown that $\mathfrak{G}$ can be partitioned into a union of disjoint closed subsets $\mathfrak{G}(x_0)$, $x_0 \in X_\infty$, where X_∞ is the one-point compactification of X. Because of (8.1), this yields a notion of weak convergence at points of X_∞ and leads to a related idea, the essential range at x_0 of $u \in L_\infty(X, \mathcal{B}, \lambda)$ (sometimes called the set of cluster values of u at x_0) which is parameterized by elements of $\mathfrak{G}(x_0)$. The pointwise criterion for weak convergence in the previous chapter can then be localised and related to the pointwise essential range.

The distinction between the essential range at x_0 of $u \in L_\infty(X, \mathcal{B}, \lambda)$ and its actual range can be rather striking. When $\Omega \subset \mathbb{R}^n$ with positive Lebesgue measure, there exists $u \in L_\infty(\Omega, \mathcal{B}, \lambda)$ for which the range is denumerable, $\{u(x) : x \in \Omega\} = \mathbb{Q} \cap [0, 1]$, while at every $x_0 \in \Omega$ the essential range of u is the continuum $[0, 1]$. The essential range of $u \in \ell_\infty(\mathbb{N})$ at the point at infinity in $\mathbb{N}_\infty$ is the set of limits of convergent subsequences $\{u(k_j)\}$ of $\{u(k)\}$.

9.1 Localising $\mathfrak{G}$

When (X, ϱ) is a locally compact Hausdorff space, by (5.1),

$$\mathfrak{G} = \Big\{ \omega \in L_\infty^*(X, \mathcal{B}, \lambda) : \ \omega(X) = 1, \ \omega(A) \in \{0, 1\}, A \in \mathcal{B} \Big\}.$$

Lemma 9.1 *Let (X, ϱ) be locally compact and Hausdorff.*
(a) For $\omega \in \mathfrak{G}$ there exists a compact set K with $\omega(K) = 1$ if and only if there exists $x_0 \in K$ such that

© The Author(s), under exclusive license to Springer Nature Switzerland AG 2020
J. Toland, *The Dual of $L_\infty(X, \mathcal{L}, \lambda)$, Finitely Additive Measures
and Weak Convergence*, SpringerBriefs in Mathematics,
https://doi.org/10.1007/978-3-030-34732-1_9

$$\omega(G) = 1 \text{ for all open sets } G \text{ with } x_0 \in G. \tag{9.1a}$$

For all $\omega \in \mathfrak{G}$ there is at most one $x_0 \in X$ satisfying (9.1a) and when (X, ϱ) is compact there is exactly one such x_0.

(b) For $x_0 \in X$ at least one $\omega \in \mathfrak{G}$ satisfies (9.1a). If (X, ϱ) is not compact, at least one $\omega \in \mathfrak{G}$ satisfies,

$$\omega(X \setminus K) = 1 \text{ for all compact } K \subset X. \tag{9.1b}$$

Proof (a) Suppose that $\omega(K) = 1$ for some compact K and (9.1a) is false. Then for each $x \in K$ there is an open set G_x with $x \in G_x$ and $\omega(G_x) = 0$. Since K is compact, $K \subset \bigcup_{i=1}^{K} G_{x_i}$. Since $\omega(G_{x_i}) = 0$, $1 \leqslant i \leqslant K$ and ω is finitely additive, $\omega(K) = 0$ which is false. Hence if $\omega(K) = 1$ and K is compact, there exists $x_0 \in K$ satisfying (9.1a). Suppose there is another $x_1 \in X$ satisfying (9.1a). Since X is Hausdorff, there exist open sets with $x_0 \in G_{x_0}$, $x_1 \in G_{x_1}$ and $G_{x_0} \cap G_{x_1} = \emptyset$. But this is impossible because, by finite additivity, it would imply that $\omega(G_{x_0} \cup G_{x_1}) = 2$.

Now suppose $\omega \in \mathfrak{G}$ and $\omega(K) = 0$ for all compact sets K. By local compactness, for any $x \in X$ there is an open set G_x with $x \in G_x$ and its closure $\overline{G_x}$ is compact. Hence, since $\omega(G_x) \leqslant \omega(\overline{G_x}) = 0$, there is no $x \in X$ with the required property. Finally, the existence of x_0 when X is compact follows because $\omega(X) = 1$.

(b) For $x_0 \in X$, let $\mathcal{E}(x_0) = \{G \in \varrho : x_0 \in G\}$. Since by hypothesis $\lambda(G) > 0$ when G is non-empty and open, by Theorem 5.6 there exists $\omega \in \mathfrak{G}$ with $\omega(G) = 1$ for all $G \in \mathcal{E}(x_0)$.

Finally if (X, ϱ) is not compact let $\mathcal{E}_\infty = \{X \setminus K : K \text{ compact}\}$. Since sets in $\mathcal{E}_\infty$ are non-empty and open (because compact sets in Hausdorff spaces are closed), the existence of $\omega \in \mathfrak{G}$ follows from Theorem 5.6 as before. This completes the proof. $\square$

For $\omega \in \mathfrak{G}$, let ω_∞ be defined on Borel subsets E of X_∞, the one-point compactification of X (Definition 2.32), by $\omega_\infty(E) = \omega(E \cap X)$. Then ω_∞ is the unique finitely additive measure on X_∞ which takes only values 0 and 1 and which coincides with ω on Borel sets in X. Lemma 9.1 can then be re-stated as follows.

Corollary 9.2 *(a) For $\omega \in \mathfrak{G}$ there is a unique $x_0 \in X_\infty$ such that $\omega_\infty(G) = 1$ when $x_0 \in G$ and G is open in X_∞. Moreover $x_0 = x_\infty$ if and only if $\omega(K) = 0$ for all compact $K \subset X$.*

(b) For $x_0 \in X_\infty$ there exists $\omega \in \mathfrak{G}$ such that $\omega(G) = 1$ when $x_0 \in G$ and G is open in X_∞.

Definition 9.3 For $x_0 \in X_\infty$, let $\mathfrak{G}(x_0) \subset \mathfrak{G}$ denote the set of $\omega \in \mathfrak{G}$ for which the conclusion of Corollary 9.2(b) holds, and let $\mathfrak{U}(x_0) \subset \mathfrak{U}$ denote the corresponding family of ultrafilters (Definition 5.3). $\square$

Properties of $\mathfrak{G}(x_0)$, $x_0 \in X_\infty$

(a) $\mathfrak{G}(x_0) \neq \emptyset$ by Corollary 9.2(b);

(b) $\mathfrak{G}(x_1) \cap \mathfrak{G}(x_2) = \emptyset$ if $x_1 \neq x_2 \in X_\infty$, because $(X_\infty, \varrho_\infty)$ is Hausdorff;

(c) $\mathfrak{G}(x_0)$ is compact because $(\mathfrak{G}, \tau)$ is compact by Theorem 7.3 and

$$\mathfrak{G}(x_0) = \bigcap_{\{G : x_0 \in G \in \varrho_\infty\}} \Delta_G, \text{ where } \Delta_G \text{ is closed.}$$

Theorem 9.4 *Let* (X, ϱ) *be locally compact and Hausdorff.*

(a) Suppose $x_0 \in X$ *and* $\{G_k\}_{k \in \mathbb{N}}$ *is a nested, denumerable local base for* ϱ *at* x_0 *(i.e.* $x_0 \in G_{k+1} \subset G_k \in \varrho$ *for all k and* $x_0 \in G \in \varrho$ *implies* $G_k \subset G$ *for some k). Then*

(i) $\mathfrak{G}(x_0)$ *is a compact* G_δ-set in $(\mathfrak{G}, \tau)$;
(ii) $\mathfrak{G}(x_0)^\circ \neq \emptyset$ *if and only if* $\{x_0\}$ *is an atom and* $\mathfrak{G}(x_0)^\circ = \{\delta_{x_0}\}$, *a Dirac measure;*
(iii) $\mathfrak{G}(x_0)^\circ = \emptyset$ *and* $\mathfrak{G}(x_0)$ *is uncountable if* $\{x_0\}$ *is not an atom.*

(b) Suppose (X, ϱ) *is not compact and* $\{\{x_\infty\} \cup (X \setminus K_k)\}_{k \in \mathbb{N}}$ *is a nested, denumerable local base for* ϱ_∞ *at* x_∞ *(i.e.* K_k *is compact,* $K_k \subset K_{k+1}$ *and for every compact K there exists $k \in \mathbb{N}$ with* $K \subset K_k$). *Then*

(i) $\mathfrak{G}(x_\infty)$ *is a closed* G_δ-set in $(\mathfrak{G}, \tau)$,
(ii) $\mathfrak{G}(x_\infty)^\circ = \emptyset$ *and* $\mathfrak{G}(x_\infty)$ *is uncountable.*

Proof (a) (i) For $x_0 \in X$ it is immediate that $\mathfrak{G}(x_0) = \bigcap_k \Delta_{G_k}$ and since Δ_{G_k} is both open and closed, $\mathfrak{G}(x_0)$ is a closed G_δ-set in $(\mathfrak{G}, \tau)$, and $(\mathfrak{G}, \tau)$ is compact.

(ii) Since (X, ϱ) is Hausdorff, $\bigcap_{k \in \mathbb{N}} G_k = \{x_0\}$ and by Theorem 7.10 the interior $\mathfrak{G}(x_0)^\circ = \Delta_{\{x_0\}}$. Hence $\mathfrak{G}(x_0)^\circ \neq \emptyset$ if and only if $\lambda(\{x_0\}) > 0$, i.e. $\{x_0\}$ is an atom and by Lemma 7.7 $\mathfrak{G}(x_0)^\circ$ is the singleton $\Delta_{\{x_0\}}$.

(iii) Since $x_0 \in G_k$, by hypothesis $\lambda(G_k) > 0$ because G_k is non-empty and open, and $\lambda(G_k) \to 0$, since λ is σ-additive and $\lambda(\{x_0\}) = 0$. That $\mathfrak{G}(x_0)$ is uncountable and $\mathfrak{G}(x_0)^\circ = \emptyset$ now follows from Theorem 7.10.

(b) Since a singleton is compact, the hypothesis implies that $X = \bigcup_{k \in \mathbb{N}} K_k$, and since $\mathfrak{G}(x_\infty) = \bigcap_k \Delta_{(X \setminus K_k)}$ the results follow from Theorem 7.10 as in part (a). $\quad\square$

The hypothesis of part (b) implies that $X = \bigcup_k K_k$; in other words, (X, ϱ) is σ-compact (Definition 2.31). However, (X, ϱ) being σ-compact does not imply that the hypotheses of part (b) are satisfied.

Remark 9.5 Given the one-to-one correspondence between the set $\mathfrak{G}$ of 0–1 finitely additive measures and ultrafilters $\mathfrak{U}$ in $(X, \mathcal{B}, \lambda)$ in Theorem 5.4, there is an obvious similarity with Lemma 7.16, but there are important differences. By Lemma 7.16 there is a one-to-one correspondence between points $x_0 \in X$ and Dirac measures $\delta_{x_0} \in \mathfrak{D}$, and no Dirac measure is concentrated at infinity. By contrast, there is a one-to-one correspondence between points $x_0 \in X_\infty$ and the disjoint sets $\mathfrak{G}(x_0) \subset \mathfrak{G}$ which by Theorem 9.4(iii) may be uncountably infinite. $\quad\square$

9.2   Localising Weak Convergence

In the notation of Definition 9.3

$$\mathfrak{G} = \bigcup_{x_0 \in X_\infty} \mathfrak{G}(x_0), \qquad \mathfrak{U} = \bigcup_{x_0 \in X_\infty} \mathfrak{U}(x_0), \tag{9.2}$$

which leads to the following definition of pointwise weak convergence.

Definition 9.6 (*Weak Convergence at a Point*) A bounded sequence $\{u_k\}$ in $L_\infty(X, \mathcal{B}, \lambda)$ converges weakly to u at $x_0 \in X_\infty$ (written as $u_k \rightharpoonup u$ at x_0) if and only if

$$\int_X u_k \, d\omega \to \int_X u \, d\omega \text{ as } k \to \infty \text{ for all } \omega \in \mathfrak{G}(x_0).$$

Equivalently, for all open sets G with $x_0 \in G$,

$$\int_G u_k \, d\omega \to \int_G u \, d\omega \text{ for all } \omega \in \mathfrak{G}(x_0).$$

$\square$

Theorem 9.7 *A bounded sequence converges weakly to u in $L_\infty(X, \mathcal{B}, \lambda)$ if and only if it converges weakly to u at every $x_0 \in X_\infty$.*

Proof This is immediate from (8.1), (9.2) and Definition 9.6. $\square$

For $u \in L_\infty(X, \mathcal{B}, \lambda)$, $\alpha > 0$ and G open in X_∞, let

$$A_\alpha(u|_G) = \left\{ x \in G \bigcap X : |u(x)| > \alpha \right\}.$$

Theorem 9.8 *For bounded $\{u_k\}$ in $L_\infty(X, \mathcal{B}, \lambda)$ the following are equivalent:*

(a) $u_k \not\rightharpoonup 0$ at $x_0 \in X_\infty$;
(b) *there exists $\alpha_0 > 0$ and a strictly increasing sequence $\{k_j\} \subset \mathbb{N}$ such that for all open $G \subset X_\infty$ with $x_0 \in G$ and every $J \in \mathbb{N}$*

$$\lambda\left(\bigcap_{j=1}^{J} A_{\alpha_0}(u_{k_j}|_G) \right) > 0; \tag{9.3}$$

(c) *there exists a strictly increasing $\{k_j\}$ such that, for all open $G \subset X_\infty$ with $x_0 \in G$, the non-increasing sequence $\{v_J\}$ defined by (8.4) satisfies*

$$v_J|_{G \bigcap X} \not\rightharpoonup 0 \text{ in } L_\infty(G \bigcap X, \mathcal{B}, \lambda) \text{ as } J \to \infty.$$

Proof That (b) and (c) are equivalent is immediate from the definitions. If (a) holds there exists $\alpha_0 > 0$, $\omega_0 \in \mathfrak{G}(x_0)$ and a strictly increasing sequence $\{k_j\} \subset \mathbb{N}$, such that for any open G with $x_0 \in G$

$$0 < \alpha_0 \leqslant \int_X |u_{k_j}| \, d\omega_0 = \int_G |u_{k_j}| \, d\omega_0 \text{ for all } j.$$

Now (b) follow as in the proof of Theorem 8.7. If (b) holds then (9.3) implies that

$$\mathcal{E}_0 := \left\{ A_\alpha(u_{k_j}|_G) : j \in \mathbb{N}, \ G \text{ open}, \ x_0 \in G \right\}$$

satisfies the hypothesis of Theorem 5.6. Hence, there exists $\omega_0 \in \mathfrak{G}$ with $\omega_0(A) = 1$ for all $A \in \mathcal{E}_0$. It follows that $\omega_0 \in \mathfrak{G}(x_0)$ and from (9.3),

$$\int_G |u_{k_j}| \, d\omega_0 = \int_X |u_{k_j}| \, d\omega_0 \geqslant \alpha_0 > 0 \text{ for all } j,$$

and (a) follows. $\square$

Example 9.9 Suppose $(X, \mathcal{B}, \lambda)$ has no atoms and (X, ϱ) is completely separable (i.e. ϱ has a denumerable base, $\mathcal{G}$ say, see Sect. 2.10). By Theorem 2.22 there exists $\{A_k\}$, a sequence of mutually disjoints sets in $\mathcal{B}$, each of which splits $\mathcal{G}$.

Hence, for all $k \in \mathbb{N}$ and for every open set $G \subset X$ the function χ_{A_k} takes the value 1 on a subset of G with positive measure. However, since the sets A_k are mutually disjoint it follows from Example 8.14 that $\chi_{A_k} \rightharpoonup 0$ in $L_\infty(X, \mathcal{B}, \lambda)$, and therefore $\chi_{A_k} \rightharpoonup 0$ at every point of X. Note that $\{\chi_{A_{k_j}}\}$ does not satisfy (9.3) for any $\{k_j\} \subset \mathbb{N}$ and $x_0 \in X$. $\square$

9.3 Fine Structure at x_0 of $u \in L_\infty(X, \mathcal{B}, \lambda)$

By analogy with (6.11), in a Borel measure space $(X, \mathcal{B}, \lambda)$ the essential range of u localised at $x_0 \in X_\infty$ is defined as

$$\mathcal{R}(u)(x_0) = \bigcap_{\{\epsilon > 0, \ G : x_0 \in G \in \varrho\}} \left\{ \alpha : \lambda \left(\left\{ x \in G \bigcap X : |\alpha - u(x)| < \epsilon \right\} \right) > 0 \right\} \quad (9.4)$$

$$= \left\{ \int_X u \, d\omega : \omega \in \mathfrak{G}(x_0) \right\}.$$

From the first line of (9.4), the multivalued mapping $x_0 \to \mathcal{R}(u)(x_0)$ reflects the measure-theoretic fine structure at x_0 of $u \in L_\infty(X, \mathcal{B}, \lambda)$ and, since

$$\mathcal{R}(u)(x_0) = L[u](\mathfrak{G}(x_0)), \quad (9.5)$$

the second line relates that fine structure to the isometric isomorphism $L :$ $L_\infty(X, \mathcal{B}, \lambda) \to C(\mathfrak{G}, \tau)$ in Sect. 7.2. $\mathcal{R}(u)(x_0)$ is closed because $\mathfrak{G}(x_0)$ is compact and $L[u]$ is continuous.

A point $\alpha \in \mathcal{R}(u)(x_0)$ given by $\alpha = \int_X u\, d\omega$, $\omega \in \mathfrak{G}(x_0)$, may be thought of as the directional limit of u at x_0, the "direction" being determined by the ultrafilter $\mathcal{U}_\omega \in \mathfrak{U}(x_0)$ (see (5.3b)). Thus, by Theorem 9.7, weak convergence in $L_\infty(X, \mathcal{B}, \lambda)$ is equivalent to convergence, for each $\mathcal{U} \in \mathfrak{U}(x_0)$, of the directional limits of u_k at x_0 to corresponding directional limits of u at x_0, for each $x_0 \in X_\infty$. In other words, $u_k \rightharpoonup u$ implies that for every $x_0 \in X_\infty$ and every $\alpha \in \mathcal{R}(u)(x_0)$ there exist $\alpha_k \in \mathcal{R}(u_k)(x_0)$ such that $\alpha_k \to \alpha$ as $k \to \infty$. This is not equivalent to $\alpha_k \to \alpha$ when $\alpha_k \in \mathcal{R}(u_k)(x_0)$ and $\alpha \in \mathcal{R}(u)(x_0)$ because of the possibility that

$$\alpha_k = \int_X u_k\, d\omega_k \text{ and } \alpha = \int_X u\, d\omega, \text{ but } \omega_k \neq \omega.$$

Thus the multivalued $\mathcal{R}(u)(x_0)$ may be interpreted as representing the fine structure at x_0 of $u \in L_\infty(X, \mathcal{B}, \lambda)$ which is intimately related to weak convergence. A sufficient condition for $u_k \rightharpoonup u$ is that for every $x_0 \in X_\infty$

$$\sup\left\{ |\alpha| :\ \alpha \in \mathcal{R}(u_k - u)(x_0) \right\} \to 0 \text{ as } k \to \infty.$$

As already noted, the necessary condition is not sufficient. The following example shows that the sufficient condition is not necessary.

Example 9.10 With $\{A_k\}$ in Example 9.9, $u_k = \chi_{A_k}$ has the property that $\mathcal{R}(u_k)(x) = \{0, 1\}$ for all $x \in X$ while $u_k \rightharpoonup 0$ in $L_\infty(X, \mathcal{B}, \lambda)$ by Corollary 3.2. For a simpler example, let $u_k = \chi_{A_k}$ where $\{A_k\}$ is a sequence of disjoint open segments centred on the origin 0 of the unit disc in $\mathbb{R}^2$. Then $\mathcal{R}(u_k)(0) = \{0, 1\}$ but $u_k \rightharpoonup 0$ by Corollary 3.2. In these examples $\int_X u_k\, d\omega \to 0$, but not uniformly for $\omega \in \mathfrak{G}(0)$. □

The next example illustrates the strong distinction between the pointwise value $u(x)$ and the essential range $\mathcal{R}(u)(x)$ at a point x.

Example 9.11 Suppose $(X, \mathcal{B}, \lambda)$ has no atoms and (X, ϱ) is completely separable with denumerable base $\mathcal{G}$. Let $\{A_i : i \in \mathbb{N}\}$ be the disjoint family of measurable sets, each of which splits $\mathcal{G}$ according to Theorem 2.25. Let $q : \mathbb{N} \to \mathbb{Q} \cap (0, 1)$ be a bijection and define a measurable function by

$$u(x) = \sum_{i=1}^\infty q(i)\chi_{A_i} = \left\{ \begin{matrix} q(i) \text{ if } x \in A_i,\ i \in \mathbb{N} \\ 0 \text{ otherwise} \end{matrix} \right\}.$$

From Theorem 2.25 and (9.4) it follows that

$$\mathbb{Q} \cap (0, 1) \subset \mathcal{R}(u)(x_0) \subset [0, 1] \text{ for all } x_0 \in X.$$

Since $\mathcal{R}(u)(x_0)$ is closed, $\mathcal{R}(u)(x_0) = [0, 1]$, even though $u(x_0) \in \mathbb{Q}$, for all $x_0 \in X$. $\square$

Lemma 9.12 *Suppose (X, ϱ) is completely separable and $u \in L_\infty(X, \mathcal{B}, \lambda)$. Then $u(x) \in \mathcal{R}(u)(x)$ for λ-almost all $x \in X$.*

Proof Consider first the case when u is simple, say $u = \sum_{k=1}^{K} \alpha_k \chi_{A_k}$ where the α_k are distinct and $A_k = \{x : u(x) = \alpha_k\} \in \mathcal{B}$. Then let

$$B_k = \left\{ x \in X : u(x) = \alpha_k \notin \mathcal{R}(u)(x) \right\}, \quad k \in \{1, \cdots, K\},$$

$$\text{so that } \left\{ x \in X : u(x) \notin \mathcal{R}(u)(x) \right\} = \bigcup_{k=1}^{K} B_k.$$

Let $y \in B_k$. Then $u(y) = \alpha_k$ and there exists an open set G with $y \in G$ and $\lambda(\{x \in G : u(x) = \alpha_k\}) = 0$. Since there is no loss in assuming that $G \in \mathcal{G}$ where $\mathcal{G}$ is the denumerable base for ϱ, B_k is covered by a denumerable family of sets of zero measure. Hence $\lambda(B_k) = 0$ for all k, and so $\lambda(\{x : u(x) \notin \mathcal{R}(u)(x)\}) = 0$.

For the general case, let $u \in L_\infty(X, \mathcal{B}, \lambda)$ be arbitrary, let $\{u_k\}$ be a sequence of simple functions with $\|u_k - u\|_\infty \to 0$ in $L_\infty(X, \mathcal{B}, \lambda)$ and for fixed k let $u_k(x) \in \mathcal{R}(u_k)(x)$ for all $x \in E_k$ where $\lambda(X \setminus E_k) = 0$. Then $\lambda(X \setminus E) = 0$ where $E = \bigcap_{k \in \mathbb{N}} E_k$,

$$u_k(x) \in \mathcal{R}(u_k)(x) \text{ for all } k \in \mathbb{N} \text{ and } x \in E,$$

and without loss of generality suppose $u_k(x) \to u(x)$ for all $x \in E$.

Now for $x \in E$ and $k \in \mathbb{N}$ let $\omega_k \in \mathfrak{G}(x)$ be such that $u_k(x) = \int_X u_k \, d\omega_k$. Then for $x \in E$,

$$u(x) = \lim_{k \to \infty} u_k(x) = \lim_{k \to \infty} \int_X u_k \, d\omega_k$$

$$= \lim_{k \to \infty} \left(\int_X (u_k - u) \, d\omega_k + \int_X u \, d\omega_k \right) = \lim_{k \to \infty} \int_X u \, d\omega_k \in \mathcal{R}(u)(x),$$

since $\int_X u \, d\omega_k \in \mathcal{R}(u)(x)$ and $\mathcal{R}(u)(x)$ is closed. The proof is complete. $\square$

Remark 9.13 As in Example 5.8, $\ell_\infty(\mathbb{N}) = L_\infty(X, \mathcal{B}, \lambda)$ where $X = \mathbb{N}$, $\mathcal{B} = \wp(\mathbb{N})$ and λ is counting measure. Then $\mathfrak{G}(k)$, $k \in \mathbb{N}$, has only one element, the σ-additive Dirac measure δ_k (Example 2.13). Moreover, $\{E_k = k + \mathbb{N} : k \in \mathbb{N}\}$ is a local base of the open neighbourhoods of ∞ in $\mathbb{N}_\infty$. In this notation, the results of Sect. 6.5 on integrating $u \in \ell_\infty(\mathbb{N})$ are examples of the formula

$$\mathcal{R}(u)(\infty) = \left\{ \int_{\mathbb{N}} u \, d\omega : \omega \in \mathfrak{G}(\infty) \right\}, \quad u \in \ell(\mathbb{N}).$$

Thus when $u(j) = u_j \in \ell_\infty$, the essential range of u at $k \in \mathbb{N}$ is the singleton $\{u_k\}$, while at infinity the essential range is a singleton $\{\alpha\}$ if and only if $u_j \to \alpha$ as $j \to \infty$.

Lorentz [24] proved (see [31] for an elementary proof) that a sequence $\{x_k\} \in \ell_\infty$ has the property that all its Banach limits coincide and equal s if and only if as $n \to \infty$

$$y_n \to y_0 \text{ in } \ell_\infty(\mathbb{N}), \text{ where } y_n(j) = \frac{1}{n} \sum_{i=1}^{n} x_{i+j} \text{ and } y_0(j) = s \text{ for all } j.$$

This does not imply that $\lim_{k \to \infty} x_k$ exists.   □

9.4   A Localised Range from Complex Function Theory

The description of a localised essential range in (9.4) is reminiscent of the cluster set $C_D(f, z_0)$ [10] of an analytic functions f at a boundary point z_0 of its domain of definition. In Corollary 9.15, Shargorodsky [29] shows that the phenomenon illustrated in Example 9.11 arises naturally in the theory of complex Hardy spaces. Let $\mathbb{T}$ and $\mathbb{D}$ denote the unit circle and the unit disc, respectively, and let

$$B_\rho(\alpha) := \left\{ z \in \mathbb{C} : |z - \alpha| < \rho \right\}, \quad \alpha \in \mathbb{C}, \ \rho > 0.$$

Let $f \in H^\infty(\mathbb{D})$. We say that $\zeta \in \mathbb{T}$ is a singularity of f if f cannot be extended analytically to ζ (see, e.g. [17, Chap. II, Sect. 6]).

At almost every $\zeta \in \mathbb{T}$, f has a non-tangential limit, denoted by $f(\zeta)$, at ζ. The following notation differs from that of [17, Chap. II, Sect. 6]:

$$\mathcal{R}(f)(\vartheta_0) := \left\{ \alpha \in \mathbb{C} : \lambda\left(\left\{ \vartheta \in \mathbb{R} : |\vartheta - \vartheta_0| < \delta, \ \left| f(e^{i\vartheta}) - \alpha \right| < \varepsilon \right\} \right) > 0 \right.$$

$$\left. \text{for all } \varepsilon, \delta > 0 \right\}, \quad \vartheta_0 \in (-\pi, \pi],$$

where λ denotes the standard Lebesgue measure on $\mathbb{R}$.

Theorem 9.14 *Let $f \in H^\infty(\mathbb{D})$ be an inner function and let $e^{i\vartheta_0} \in \mathbb{T}$ be its singularity. Then $\mathcal{R}(f)(\vartheta_0) = \mathbb{T}$.*

Proof Since $|f| = 1$ almost everywhere on $\mathbb{T}$, one has $\mathcal{R}(f)(\vartheta_0) \subseteq \mathbb{T}$.

Suppose $\alpha \in \mathbb{T}$ does not belong to $\mathcal{R}(f)(\vartheta_0)$. Then there exist $\varepsilon, \delta \in (0, 1)$ with

$$\lambda\left(\left\{ \vartheta \in \mathbb{R} : |\vartheta - \vartheta_0| < \delta, \ \left| f(e^{i\vartheta}) - \alpha \right| < \varepsilon \right\} \right) = 0. \tag{9.6}$$

Let $\alpha_{\pm\varepsilon}$ be the two points where the circle $\{ \zeta \in \mathbb{C} : |\zeta - \alpha| = \varepsilon \}$ intersects $\mathbb{T}$ and let $\mathbb{T}(\alpha, \varepsilon)$ and $\mathbf{D}(\alpha, \varepsilon)$ be the intersections of $\mathbb{T}$ and $\mathbb{D} \bigcup \mathbb{T}$ with the closed half-plane containing 0 and bounded by the straight line through $\alpha_{\pm\varepsilon}$:

$$\mathbb{T}(\alpha, \varepsilon) := \left\{ \zeta \in \mathbb{T} \mid \operatorname{Re}\left(\alpha\overline{\zeta}\right) \leq \operatorname{Re}\left(\alpha\overline{\alpha_{\pm\varepsilon}}\right) \right\},$$

$$\mathbf{D}(\alpha, \varepsilon) := \left\{ \zeta \in \mathbb{D} \bigcup \mathbb{T} \mid \operatorname{Re}\left(\alpha\overline{\zeta}\right) \leq \operatorname{Re}\left(\alpha\overline{\alpha_{\pm\varepsilon}}\right) \right\}.$$

Let $\varrho > 0$ be the distance from α to $\mathbf{D}(\alpha, \varepsilon)$. (It is not difficult to see that ϱ is the distance from α to the midpoint of the chord $[\alpha_{-\varepsilon}, \alpha_\varepsilon]$ and $\varrho = 1 - \cos(2 \arcsin \frac{\xi}{2})$.) For the Poisson kernel

$$P_r(\vartheta) := \frac{1 - r^2}{1 - 2r \cos \vartheta + r^2}, \quad 0 \leq r < 1, \ \vartheta \in \mathbb{R},$$

there exists $\delta_0 > 0$ such that

$$\int_{\delta \leq |\vartheta - \vartheta_0| \leq \pi} P_r(\theta - \vartheta) \, d\vartheta < \varrho/2, \quad \text{for all } z = re^{i\theta} \in B_{\delta_0}(e^{i\vartheta_0}). \tag{9.7}$$

It follows from (9.6) that $f\left(e^{i\vartheta}\right) \in \mathbb{T}(\alpha, \varepsilon)$ for almost all ϑ with $|\vartheta - \vartheta_0| < \delta$. Since $P_r \geq 0$,

$$\int_{|\vartheta - \vartheta_0| < \delta} P_r(\theta - \vartheta) \, d\vartheta < \int_{|\vartheta - \vartheta_0| \leq \pi} P_r(\theta - \vartheta) \, d\vartheta = 1,$$

$\mathbb{T}(\alpha, \varepsilon) \subset \mathbf{D}(\alpha, \varepsilon)$ and the latter is a closed convex set containing 0, it follows that

$$\int_{|\vartheta - \vartheta_0| < \delta} P_r(\theta - \vartheta) f(e^{i\vartheta}) \, d\vartheta \in \mathbf{D}(\alpha, \varepsilon). \tag{9.8}$$

On the other hand, it follows from (9.7) that

$$\left| \int_{\delta \leq |\vartheta - \vartheta_0| \leq \pi} P_r(\theta - \vartheta) f(e^{i\vartheta}) \, d\vartheta \right| \leq \int_{\delta \leq |\vartheta - \vartheta_0| \leq \pi} P_r(\theta - \vartheta) \, d\vartheta < \varrho/2 \tag{9.9}$$

for all $re^{i\theta} \in B_{\delta_0}(e^{i\vartheta_0})$. Then it follows from the definition of ϱ and from (9.8), (9.9) that for $z = re^{i\theta} \in B_{\delta_0}(e^{i\vartheta_0})$,

$$f(z) = \int_{|\vartheta - \vartheta_0| \leq \pi} P_r(\theta - \vartheta) f(e^{i\vartheta}) \, d\vartheta$$

$$= \int_{|\vartheta - \vartheta_0| < \delta} P_r(\theta - \vartheta) f(e^{i\vartheta}) \, d\vartheta + \int_{\delta \leq |\vartheta - \vartheta_0| \leq \pi} P_r(\theta - \vartheta) f(e^{i\vartheta}) \, d\vartheta,$$

cannot belong to $B_{\varrho/2}(\alpha)$. Since this contradicts [17, Chap. II, Sect. 6, Theorem 6.6], every $\alpha \in \mathbb{T}$ must belong to $\mathcal{R}(f)(\vartheta_0)$. So, $\mathbb{T} \subseteq \mathcal{R}(f)(\vartheta_0) \subseteq \mathbb{T}$, i.e. $\mathcal{R}(f)(\vartheta_0) = \mathbb{T}$. $\qquad \square$

If f is an infinite Blaschke product, such that the closure of its zero set contains $\mathbb{T}$, or the singular inner function defined by a singular measure with closed support equal to $\mathbb{T}$, then the set of its singularities equals $\mathbb{T}$ (see [17, Chap. II, Sect. 6, the paragraph after Theorem 6.6]).

Corollary 9.15 *Let $f \in H^\infty(\mathbb{D})$ be an inner function whose set of singularities equals $\mathbb{T}$. Then $\mathcal{R}(f)(\vartheta) = \mathbb{T}$ for every $e^{i\vartheta} \in \mathbb{T}$.*

Remark 9.16 Corollary 9.15 provides an example of a function w_1 defined on $(-\pi, \pi)$ and such that $\mathcal{R}(w_1)(x) = \mathbb{T}$ for every $x \in (-\pi, \pi)$. Let $w(x) := w_0(x_1)$, $x = (x_1, \ldots, x_n) \in (-\pi, \pi)^n$. It is easy to see that $\mathcal{R}(w)(x) = \mathbb{T}$ for every $x \in (-\pi, \pi)^n$.   □

Chapter 10
Reconciling Representations

Theorem 3.1 identifies elements of $L_\infty(X, \mathcal{L}, \lambda)^*$ with finitely additive measures in $L_\infty^*(X, \mathcal{L}, \lambda)$ and Theorem 2.37 (Riesz) identifies bounded linear functionals on $C(\mathfrak{G}, \tau)$ with regular real Borel measures, denoted by $C^*(\mathfrak{G}, \tau)$, on $(\mathfrak{G}, \tau)$. However, the isometric isomorphism in Theorem 7.4 means that these representations must be related to one another and the finite additivity of one should be reconciled with the σ-additivity of the other. This issue will be addressed first.

Then, in a topological space setting, for a functional $f \in L_\infty(X, \mathcal{B}, \lambda)^*$ let $\nu \geqslant 0$ be the finitely additive measure in the Yosida–Hewitt representation of f and $\hat{\nu} \geqslant 0$ the regular real Borel measure in the classical Riesz representation of f restricted to the space $C_0(X, \varrho)$ of continuous functions that vanish at infinity. A minimax formula for $\hat{\nu}$ in terms ν is obtained in Sect. 10.2 when ν is non-negative.

10.1 $(\mathfrak{G}, \tau)$ Versus $L_\infty(X, \mathcal{L}, \lambda)$

In this section, $(X, \mathcal{L}, \lambda)$ is the general measure space introduced in Chap. 3. For a bounded linear functional f on $L_\infty(X, \mathcal{L}, \lambda)$, let $\tilde{f} \in C(\mathfrak{G}, \tau)^*$ be defined by $\tilde{f} \circ L = f$, where L is as in Theorem 7.4. Then with $\nu \in L_\infty^*(X, \mathcal{L}, \lambda)$ representing $f \in L_\infty(X, \mathcal{L}, \lambda)^*$ let $\tilde{\nu} \in C^*(\mathfrak{G}, \tau)$ denote the regular real Borel measure on $\mathfrak{G}$ that represents $\tilde{f} \in C(\mathfrak{G}, \tau)^*$. Since, for any $A \in \mathcal{L}$ and $\omega \in \mathfrak{G}$,

$$L[\chi_A](\omega) = \int_X \chi_A \, d\omega = \omega(A) = \chi_{\Delta_A}(\omega),$$

it follows that $\chi_{\Delta_A} = L[\chi_A]$. Hence for $A \in \mathcal{L}$ and $\nu \in L_\infty^*(X, \mathcal{L}, \lambda)$,

$$\tilde{\nu}(\Delta_A) = \int_{\mathfrak{G}} \chi_{\Delta_A} \, d\tilde{\nu} = \tilde{f}(\chi_{\Delta_A}) = \tilde{f}(L[\chi_A]) = f(\chi_A) = \int_A \chi_A \, d\nu = \nu(A).$$

Thus the key relation is

© The Author(s), under exclusive license to Springer Nature Switzerland AG 2020
J. Toland, *The Dual of $L_\infty(X, \mathcal{L}, \lambda)$, Finitely Additive Measures
and Weak Convergence*, SpringerBriefs in Mathematics,
https://doi.org/10.1007/978-3-030-34732-1_10

$$\tilde{\nu}(\Delta_A) = \nu(A) \text{ for all } A \in \mathcal{L}, \ \nu \in L_\infty^*(X, \mathcal{L}, \lambda), \qquad (10.1)$$

which is compatible with (7.1b) and the additivity properties of ν and $\tilde{\nu}$. To illustrate this suppose $\{A_i : i \in \mathbb{N}\} \subset \mathcal{L}$, $A_i \cap A_j \in \mathcal{N}$ when $j \neq j$ and $0 \leqslant \nu \in L_\infty^*(X, \mathcal{L}, \lambda)$. Then the σ-additivity of $\tilde{\nu}$, the finite additivity of ν, (7.1b) and Theorem 7.10 yield

$$\tilde{\nu}\left(\bigcup_{i=1}^\infty \Delta_{A_i}\right) = \sum_{i=1}^\infty \tilde{\nu}(\Delta_{A_i}) = \sum_{i=1}^\infty \nu(A_i)$$

$$\leqslant \nu\left(\bigcup_{i=1}^\infty A_i\right) = \tilde{\nu}\left(\Delta_{\cup_{i=1}^\infty A_i}\right) = \tilde{\nu}\left(\overline{\bigcup_{i=1}^\infty \Delta_{A_i}}\right).$$

Remark 10.1 From (10.1) it follows that $\nu \geqslant 0$ on $\mathcal{L}$ if $\tilde{\nu} \geqslant 0$ on the Borel σ-algebra in $\mathfrak{G}$. Conversely, if $\nu \geqslant 0$ on $\mathcal{L}$ then $\tilde{\nu} \geqslant 0$ by Corollary 3.4 of the Riesz representation theorem.   □

Theorem 10.2 $\gamma \in L_\infty^*(X, \mathcal{L}, \lambda)$ *is σ-additive if and only if $\tilde{\gamma} \in C^*(\mathfrak{G}, \tau)$ is zero on every closed, nowhere dense G_δ-set in $(\mathfrak{G}, \tau)$.*

Proof If $\mathcal{H} \subset \mathfrak{G}$ is a closed G_δ-set which is nowhere dense, $\mathcal{H}^\circ = \emptyset$ in $(\mathfrak{G}, \tau)$. By Corollary 7.11, there is a nested sequence of sets $A_{k+1} \subset A_k$ in $\mathcal{L}$ with $\mathcal{H} = \bigcap_{k=1}^\infty \Delta_{A_k}$ and $\Delta_{\bigcap_{k=1}^\infty A_k} = \mathcal{H}^\circ = \emptyset$. Hence $\bigcap_{k=1}^\infty A_k \in \mathcal{N}$ by Theorem 5.6.

Now suppose that $\gamma \in L_\infty^*(X, \mathcal{L}, \lambda)$ is σ-additive. Then the σ-additivity of both $\tilde{\gamma}$ and γ, and the fact that $\gamma \ll \lambda$, imply that

$$\tilde{\gamma}(\mathcal{H}) = \lim_{k \to \infty} \tilde{\gamma}(\Delta_{A_k}) = \lim_{k \to \infty} \gamma(A_k) = \gamma\left(\bigcap_{k=1}^\infty A_k\right) = 0.$$

Conversely, suppose that $\tilde{\gamma}(\mathcal{H}) = 0$ for every closed, nowhere dense G_δ-set $\mathcal{H}$ in $(\mathfrak{G}, \tau)$. Let $\{E_k\} \subset \mathcal{L}$ with $E_{k+1} \subset E_k$ and $\bigcap_{k \in \mathbb{N}} E_k = \emptyset$. Since the nested sets Δ_{E_k} are open and closed, $\bigcap_k \Delta_{E_k}$ is a closed G_δ-set which is nowhere dense since, by Corollary 7.11, its interior is empty. Therefore

$$\lim_{k \to \infty} \gamma(E_k) = \lim_{k \to \infty} \tilde{\gamma}(\Delta_{E_k}) = \tilde{\gamma}\left(\bigcap_{k=1}^\infty \Delta_{E_k}\right) = 0,$$

which, by Lemma 4.2(a), proves γ is σ-additive.   □

Definition 10.3 (*Concentration of Measures*) A real measure $\tilde{\nu}$ is concentrated on a set $\mathcal{H} \subset \mathfrak{G}$ if $\tilde{\nu}(\mathfrak{B}) = 0$ for all Borel sets $\mathfrak{B} \subset \mathfrak{G} \setminus \mathcal{H}$.   □

Theorem 10.4 $\mu \in L_\infty^*(X, \mathcal{L}, \lambda)$ *is purely finitely additive if and only if $\tilde{\mu}$ is concentrated on a closed, nowhere dense G_δ-set $\mathcal{H} \subset \mathfrak{G}$.*

Proof To begin, suppose $\mu \in L_\infty^*(X, \mathcal{L}, \lambda)$ is non-negative with $\tilde{\mu}$ concentrated on a closed, nowhere dense G_δ-set $\mathcal{H}$ and let $\gamma \in \Sigma(\mathcal{L}) \cap L_\infty^*(X, \mathcal{L}, \lambda)$ with $0 \leqslant \gamma \leqslant \mu$. Then $0 \leqslant \tilde{\gamma} \leqslant \tilde{\mu}$ by Remark 10.1 and hence $\tilde{\gamma}$ is concentrated on a closed, nowhere dense G_δ-set. It follows from Theorem 10.2 that $\tilde{\gamma}$ is zero. Hence by (10.1) γ is zero and by Definition 4.8 $\mu \geqslant 0$ is purely finitely additive. The result for general μ follows because $\tilde{\mu} = \tilde{\mu}^+ - \tilde{\mu}^-$, by (4.1c), and $\tilde{\mu}^+$ and $\tilde{\mu}^-$ are concentrated on $\mathcal{H}$ if $\tilde{\mu}$ is concentrated on $\mathcal{H}$.

Since λ is σ-finite there exists $f \in L_1(X, \mathcal{L}, \lambda)$ which is strictly positive on X. Let $\gamma_f \in \Sigma(\mathcal{L})$ be defined by $\gamma_f(E) = \int_X f \, d\lambda$, $E \in \mathcal{L}$, and suppose $0 \leqslant \mu \in L_\infty^*(X, \mathcal{L}, \lambda)$ is purely finitely additive.

By Corollary 4.12, there exists a nested sequence $E_{k+1} \subset E_k$ in $\mathcal{L}$ such that $\mu(E_k) = \mu(X)$ for all k and $\gamma_f(E_k) \to 0$ as $k \to \infty$. Since γ_f is σ-additive, $\gamma_f(E) = \lim_k \gamma_f(E_k) = 0$ where $E := \bigcap_k E_k$. Therefore $\lambda(E) = 0$ since $f > 0$ λ-almost everywhere. Moreover $\tilde{\mu}(\Delta_{E_k}) = \mu(E_k) = \mu(X) = \tilde{\mu}(\mathfrak{G})$ for all k.

Since $E \in \mathcal{N}$ it follows from Theorem 7.10 that $\mathcal{H} := \bigcap_k \Delta_{E_k}$ is a closed, nowhere dense G_δ-set in $(\mathfrak{G}, \tau)$ and $\tilde{\mu}(\mathcal{H}) = \tilde{\mu}(\mathfrak{G})$ since $\tilde{\mu}$ is σ-additive. Therefore $\mathfrak{B} \subset \mathfrak{G} \setminus \mathcal{H}$ implies

$$\tilde{\mu}(\mathfrak{G}) = \tilde{\mu}(\mathfrak{B}) + \tilde{\mu}(\mathfrak{G} \setminus \mathfrak{B}) \geqslant \tilde{\mu}(\mathfrak{B}) + \tilde{\mu}(\mathcal{H}) = \tilde{\mu}(\mathfrak{B}) + \tilde{\mu}(\mathfrak{G}),$$

and hence $\tilde{\mu}(\mathfrak{B}) = 0$, as required. The general case is immediate since $\mu = \mu^+ - \mu^-$ by (4.1c). $\qquad\square$

10.2 Restriction to $C_0(X, \varrho)$ of Elements of $L_\infty^*(X, \mathcal{B}, \lambda)$

When (X, ϱ) is a locally compact Hausdorff space, a bounded linear functional on $L_\infty(X, \mathcal{B}, \lambda)$, when restricted to the Banach space $C_0(X, \varrho)$, has a representation as a regular real Borel measure $\hat{\nu}$ on X. This section explores the relationship between ν and $\hat{\nu}$. To be precise, by Corollary 3.4 for $\nu \in L_\infty^*(X, \mathcal{B}, \lambda)$, there is a unique regular real Borel measure $\hat{\nu}$ with

$$\int_X v \, d\nu = \int_X v \, d\hat{\nu} \text{ for all } v \in C_0(X, \varrho). \tag{10.2}$$

When X is compact it follows that $\nu(X) = \hat{\nu}(X)$, because in that case non-zero constant functions belong to $C_0(X, \varrho)$. However, when (X, ϱ) is not compact $\hat{\nu}(X)$ may be zero when ν is non-negative and $\nu(X) = 1$.

Since $\nu = \nu^+ - \nu^-$, without loss of generality attention will henceforth be restricted to non-negative $\nu \in L_\infty^*(X, \mathcal{B}, \lambda)$ with the goal of understanding how $\hat{\nu} \geqslant 0$ depends on $\nu \geqslant 0$. Note:

(i) from the Yosida–Hewitt decomposition (6.9), $\nu = \mu + g\lambda$ where $\mu \in L_\infty^*(X, \mathcal{B}, \lambda)$ is purely finitely additive and $g\lambda$, $g \in L_1(X, \mathcal{B}, \lambda)$, is σ-additive.

(ii) from Theorem 2.28 (Lebesgue–Radon–Nikodym), $\hat{\nu} = \rho + k\lambda$ where ρ and $k\lambda$, $k \in L_1(X, \mathcal{B}, \lambda)$, are σ-additive, ρ singular and $k\lambda$ is absolutely continuous with respect to λ (see Definition 2.18).

As a consequence, for $v \in C_0(X, \varrho)$,

$$\int_X v\, d\mu + \int_X vg\, d\lambda = \int_X v\, d\nu = \int_X v\, d\hat{\nu} = \int_X v\, d\rho + \int_X vk\, d\lambda. \qquad (10.3a)$$

As Valadier and Hensgen noted (see Sect. 6.6), the relations between μ and ρ, and g and k are not straightforward: for Lebesgue measure λ on $[0, 1]$, there is a non-negative $\nu \in \Pi(\mathcal{B})$ with

$$\int_0^1 v\, d\nu = \int_0^1 v\, d\lambda \text{ for all } v \in C[0, 1], \qquad (10.3b)$$

which means $0 \neq \mu \in \Pi(\mathcal{B})$ and $g = 0$ but $\rho = 0$ and $k \equiv 1$. Subsequently, Abramovich and Wickstead [1] provided wide a ranging generalisation of this observation. Recently, Wrobel [34] gave a sufficient condition on a purely finitely additive ν on $L_\infty[0, 1]$ for $\hat{\nu}$ to be singular with respect to Lebesgue measure. His result is essentially a special case of the "if" part of Corollary 10.10 when X is compact.

To find a formula for $\hat{\nu}$ satisfying (10.2) for a given non-negative $\nu \in L_\infty^*(X, \mathcal{B}, \lambda)$, and to characterise those ν for which $\hat{\nu}$ has a singularity, recall the following version of Urysohn's Lemma 2.61 .

Lemma 10.5 ([28, Sect. 2.12]) *If X is a locally compact Hausdorff space and $K \subset G \subset X$ where K is compact and G is open, there exists a continuous function $f : X \to [0, 1]$ such that $f(K) = 1$ and $\overline{\{x : f(x) > 0\}}$ is a compact subset of G. In particular, $f \in C_0(X, \varrho)$.*

Lemma 10.6 *Suppose (X, ϱ) is a locally compact Hausdorff space, $0 \leqslant \nu \in L_\infty^*(X, \mathcal{B}, \lambda)$ and $B \in \mathcal{B}$. Then for compact K and open G with $K \subset B \subset G$,*

$$\nu(K) \leqslant \hat{\nu}(B) \leqslant \nu(G),$$
$$\nu(K) \leqslant \hat{\nu}(K) \text{ and } \hat{\nu}(G) \leqslant \nu(G) \text{ for compact } K \text{ and open } G,$$
$$\text{and } \nu(F) \leqslant \hat{\nu}(F) + \nu(X) - \hat{\nu}(X) \text{ for closed } F.$$

In particular, $\hat{\nu}(X) = \nu(X)$ implies $\nu(F) \leqslant \hat{\nu}(F)$ for all closed $F \subset X$. ($\nu(X) = \hat{\nu}(X)$ when (X, ϱ) is compact was noted following (10.2).)

Proof For a given Borel set B and $K \subset B \subset G$ as in the statement, let f be the continuous function determined in Lemma 10.5 by K and G. Then

$$\nu(K) \leqslant \int_X f\, d\nu \leqslant \nu(G) \text{ and } \hat{\nu}(K) \leqslant \int_X f\, d\hat{\nu} \leqslant \hat{\nu}(G).$$

From (10.2), it follows that $\nu(K) \leqslant \hat{\nu}(G)$ and $\hat{\nu}(K) \leqslant \nu(G)$ whence, since $\hat{\nu}$ is a regular real Borel measure (Corollary 3.4), $\nu(K) \leqslant \hat{\nu}(B) \leqslant \nu(G)$. In particular, if $B = K$ is compact, $\nu(K) \leqslant \hat{\nu}(K)$, and if $B = G$ is open, $\hat{\nu}(G) \leqslant \nu(G)$. That $\nu(F) \leqslant \hat{\nu}(F) + \nu(X) - \hat{\nu}(X)$ when F is closed follows by finite additivity since $X \setminus F$ is open and $0 \leqslant \nu(X), \hat{\nu}(X) < \infty$. $\qquad\square$

Remark 10.7 A non-negative finitely additive set function ν on $\mathcal{B}$ is said to be regular [12, III.5.11] if for all $E \in \mathcal{B}$ and $\epsilon > 0$ there are sets $F \subset E \subset G$ with F closed, G open and $\nu(G \setminus F) < \epsilon$. If X is compact and ν is regular in this sense, by a theorem of Alexandroff [2, Pt. I, p. 590], [12, III.5.13] ν is σ-additive and hence $\hat{\nu} = \nu$. By Lemma 10.6, if $\nu(X) = \hat{\nu}(X)$ and $F \subset E \subset G$, where F is closed and G is open,

$$\nu(F) \leqslant \hat{\nu}(F) \leqslant \hat{\nu}(E) \leqslant \hat{\nu}(G) \leqslant \nu(G).$$

Hence, $\nu \geqslant 0$ regular implies $\nu = \hat{\nu}$ is σ-additive on $\mathcal{B}$ if $\nu(X) = \hat{\nu}(X)$. $\qquad\square$

Theorem 10.8 *Suppose that (X, ϱ) is a locally compact Hausdorff space, $K \subset G$ where K is compact, G is open and $\nu \in L_\infty^*(X, \mathcal{B}, \lambda)$ is non-negative. Then for $n \in \mathbb{N}$ there exists a compact set K_n and an open set G_n with*

$$K \subset G_n \subset K_n \subset G, \quad G_n \subset G_{n-1}, \ K_n \subset K_{n-1},$$
$$\hat{\nu}(K) \leqslant \nu(K_n), \quad \hat{\nu}(G) \geqslant \nu(G_n) \ and \ \lambda(K_n) < \lambda(K) + 1/n.$$

Proof Since λ is a regular Borel measure that is finite on compact sets there exist open sets G^k with $K \subset G^k \subset G$ and $\lambda(G^k) < \lambda(K) + 1/k$ for $k \in \mathbb{N}$.

Since $K \subset G^k$, for $k \in \mathbb{N}$ there exists, by Lemma 10.5, a continuous function $f_k : X \to [0, 1]$ such that $f_k(K) = 1$ and $\{x : f_k(x) > 0\}$ is a compact subset of G^k. For $x \in X$, let $g_n(x) = \min\{f_k(x) : k \leqslant n\}$ so that $g_n \leqslant g_{n-1}$, g_n is continuous on X, $g_n(K) = 1$ and $\{x : g_n(x) > 0\} \subset G^n$ is compact.

Let $G_n = \{x : g_n(x) > 0\}$ and $K_n = \overline{\{x : g_n(x) > 0\}}$. Then $K \subset G_n \subset K_n \subset G^n \subset G$ and, by Lemma 10.6,

$$\hat{\nu}(K) \leqslant \hat{\nu}(G_n) \leqslant \nu(G_n) \leqslant \nu(K_n), \quad \hat{\nu}(G) \geqslant \hat{\nu}(K_n) \geqslant \nu(K_n) \geqslant \nu(G_n),$$

and $\lambda(K_n) < \lambda(K) + 1/n$ because $K_n \subset G^n$. Now $\{G_n\}$ and $\{K_n\}$ are nested sequences of open and compact sets, respectively, because $g_n(x)$ is decreasing in n, with the required properties. This completes the proof. $\qquad\square$

Corollary 10.9 *For G open, K compact and $\nu \in L_\infty^*(X, \mathcal{B}, \lambda)$ non-negative,*

$$\hat{\nu}(G) = \sup\{\nu(K) : K \subset G, K \ compact\},$$
$$\hat{\nu}(K) = \inf\{\nu(G) : K \subset G, G \ open\}.$$

Proof From Lemma 10.6, for any open set G and all compact $K \subset G$, $\nu(K) \leqslant \hat{\nu}(K) \leqslant \hat{\nu}(G)$. Since $\hat{\nu}$ is a regular real Borel measure, for any $\epsilon > 0$ there exists compact $K \subset G$ with $\hat{\nu}(K) > \hat{\nu}(G) - \epsilon$. Now by Theorem 10.8 there exists compact K_1 with $K \subset K_1 \subset G$ and $\nu(K_1) \geqslant \hat{\nu}(K) > \hat{\nu}(G) - \epsilon$. This establishes the first identity. For a given compact set K and any open set G with $K \subset G, \hat{\nu}(K) \leqslant \hat{\nu}(G) \leqslant \nu(G)$ and for $\epsilon > 0$, there exists an open G with $K \subset G$ and $\hat{\nu}(G) < \hat{\nu}(K) + \epsilon$. By Theorem 10.8, there exists an open set G_1 with $K \subset G_1 \subset G$ with $\hat{\nu}(K) + \epsilon > \hat{\nu}(G) \geqslant \nu(G_1)$, and the result follows.   □

Say that $\hat{\nu}$ has a singularity with respect to λ if $\hat{\nu}(E) \neq 0$ (equivalently $\rho(E) \neq 0$) for some $E \in \mathcal{N}$.

Corollary 10.10 *Let (X, ϱ) be a locally compact Hausdorff space and $0 \leqslant \nu \in L_\infty^*(X, \mathcal{B}, \lambda)$. Then $0 \leqslant \hat{\nu} \in \Sigma(\mathcal{B})$ has a singularity if and only if there exists $\alpha > 0$ and a sequence of compact sets with $\nu(K_n) \geqslant \alpha$, $K_{n+1} \subset K_n$ for all n, and $\lambda(K_n) \to 0$ as $n \to \infty$.*

Proof If $\alpha > 0$ and such a sequence exists, by Lemma 10.6, $\hat{\nu}(K_n) \geqslant \alpha$ for all n. Since $\{K_n\}$ is nested and $\hat{\nu}$ is σ-additive it follows that $\hat{\nu}(K) \geqslant \alpha$ where $K = \bigcap_n K_n$. Since $K \in \mathcal{N}$, because $\lim_{n\to\infty} \lambda(K_n) = 0$ and λ is σ-additive, $\hat{\nu}$ has a singularity.

Conversely, if $\hat{\nu} \geqslant 0$ has a singularity there exists $E \in \mathcal{N}$ and $\alpha > 0$ with $\hat{\nu}(E) = 2\alpha$. Since $\hat{\nu}$ is a real Borel measure which is regular (Definition 2.21), there exists a compact $K \subset E$ with $\hat{\nu}(K) \geqslant \alpha > 0$. Now since $\lambda(K) = 0$ because $K \subset E \in \mathcal{N}$, the existence of a sequence $\{K_n\}$ of compact sets with $\nu(K_n) \geqslant \hat{\nu}(K) \geqslant \alpha$, $K_{n+1} \subset K_n$ for all n, and $\lambda(K_n) \to 0$ as $n \to \infty$ follows from Theorem 10.8.   □

Theorem 10.11 *For (X, ϱ) locally compact and $0 \leqslant \nu \in L_\infty^*(X, \mathcal{B}, \lambda)$,*

$$\hat{\nu}(B) = \inf_{\substack{G \text{ open} \\ B \subset G}} \left\{ \sup_{\substack{K \text{ compact} \\ K \subset G}} \nu(K) \right\} = \sup_{\substack{K \text{ compact} \\ K \subset B}} \left\{ \inf_{\substack{G \text{ open} \\ K \subset G}} \nu(G) \right\},$$

for all $B \in \mathcal{B}$.

Proof Since $\hat{\nu}$ is a regular real Borel measure,

$$\hat{\nu}(B) = \inf\{\hat{\nu}(G) : B \subset G, \ G \text{ open}\} = \sup\{\hat{\nu}(K) : K \subset B, \ K \text{ compact}\},$$

and the formulae follow from Corollary 10.9.   □

Corollary 10.12 *For a locally compact Hausdorff space (X, ϱ) and $\omega \in \mathfrak{G}$,*

(a) either $\hat{\omega}$ is zero or $\hat{\omega}$ is a Dirac measure δ_{x_0};

(b) both possibilities may occur when (X, ϱ) is not compact;

(c) if $\hat{\omega} = \delta_{x_0} \in \mathfrak{D}$, then $\omega \in \mathfrak{G}(x_0)$, i.e.(9.1a) holds .

Proof (a) By Lemma 9.1, either $\omega(K) = 0$ for all compact K, in which case $\hat{\omega} = 0$ by the first formula for $\hat{\omega}(B)$, or $\omega(K) = 1$ for some compact K. In the latter case, there is a unique $x_0 \in X$ for which $\omega(G) = 1$ if $x_0 \in G$ and G is open. From the second formula for $\hat{\omega}(B)$ it is immediate that $\hat{\omega}(B) = 1$ if and only if $x_0 \in B$. Hence $\hat{\omega} = \delta_{x_0} \in \mathfrak{D}$.

(b) For an example of both possibilities let $X = (0, 1)$ with the standard (locally compact but not compact) topology and Lebesgue measure. Let $\omega \in \mathfrak{G}$ be defined by Theorem 5.6 with $E_\ell = (0, 1/\ell)$, $\ell \in \mathbb{N}$. Then $\omega(K) = 0$ for all compact $K \subset (0, 1)$ and hence $\hat{\omega} = 0$. On the other hand, if $E_\ell = (1/2 + 1/\ell, 1/2)$ in Theorem 5.6, $\omega \in \mathfrak{G}$ with $\omega([1/2 + 1/\ell, 1/2]) = 1$ for all ℓ and hence $\hat{\omega} = \delta_{1/2} \in \mathfrak{D}$.

(c) If $\hat{\omega} = \delta_{x_0}$ let G an open set with $x_0 \in G$. Since $\{x_0\}$ is compact, there exists $v \in C_0(X, \varrho)$ with $v(X) \subset [0, 1]$, $v(x_0) = 1$, $v(X \setminus G) = 0$ and

$$1 \geqslant \omega(G) \geqslant \int_G v \, d\omega = \int_X v \, d\omega = \int_X v \, d\hat{\omega} = v(x_0) = 1,$$

Hence, $\omega(G) = 1$ for every open set with $x_0 \in G$. □

References

1. Y.A. Abramovich, A.W. Wickstead, Singular extensions and restrictions of order continuous functionals. Hokkaido Math. J. **21**, 475–482 (1992)
2. A.D. Alexandroff, Additive set-functions in abstract spaces I, II, III. Rec. Math. (Mat. Sbornik) N.S. **8**(50), 307–348 (1940); **9**(51), 563–628 (1941); **13**(55), 169–238 (1943)
3. C.D. Aliprantis, K.C. Border, Infinite Dimensional Analysis—A Hitchiker's Guide (Springer, Berlin, 2006)
4. J.M. Ball, Weak continuity properties of mappings and semigroups. Proc. Roy. Soc. Edin. Sect. A **72**(4), 275–280 (1975)
5. S. Banach, *Théorie des Opérations Linéaires*. Monografje Matematyczne, Tom I, Z Subwehcji Fundusku Kultryy Narodowej, Warszawa, 1932; reprinted with corrections and added material, Chelsea Publishing Co. New York, 1955; Translated from the French by F. Jellett, with comments by A. Pelczyński and Cz. Bessaga. North-Holland Mathematical Library, vol. 38. North-Holland Publishing Co., Amsterdam, 1987 & Dover Books
6. K.P.S. Bhaskara Rao, M. Bhaskara Rao, *Theory of Charges - A Study of Finitely Additive Measures*, vol. 109, Pure and Applied Mathematics (Academic Press, London, 1983)
7. N.H. Bingham, Finite additivity vs countable additivity: De Finetti and Savage. J. Électron. Hist. Probab. Stat. **6**(1), 35 (2010)
8. V.I. Bogachev, *Measure Theory*, vol. 1 (Springer, Berlin, 2007)
9. A.G. Chentsov, *Finitely Additive Measures and Relaxations of Extremal Problems*. Monographs in Contemporary Mathematics (Plenum, New York, 1996)
10. E.F. Collingwood, A.J. Lohwater, *The Theory of Cluster Sets* (Cambridge University Press, Cambridge, 1966)
11. J.B. Conway, *A Course in Functional Analysis*, 2nd edn. (Springer, New York, 2007)
12. N. Dunford, J.T. Schwartz, *Linear Operators*, vol. I (Wiley Interscience, New York, 1958)
13. G. Fichtenholz, Sur les fonctions d'ensemble additives et continues. Fund. Math. **7**(1), 296–301 (1925)
14. G. Fichtenholz, L. Kantorovitch, Sur les opérations lináires dans lespace des fonctions bornés. Stud. Math. **5**, 69–98 (1934)
15. G.B. Folland, *Real Analysis*, 2nd edn. (Wiley, New York, 1999)
16. I. Fonseca, G. Leoni, *Modern Methods in the Calculus of Variations: L^p Spaces*, Springer Monographs in Mathematics (Springer, New York, 2007)
17. J.B. Garnett, *Bounded Analytic Functions* (Revised First Edition, Springer, New York, 2007)
18. J. Heinonen, *Lectures on Analysis on Metric Spaces* (Springer, New York, 2001)

© The Author(s), under exclusive license to Springer Nature Switzerland AG 2020   95
J. Toland, *The Dual of $L_\infty(X, \mathcal{L}, \lambda)$, Finitely Additive Measures
and Weak Convergence*, SpringerBriefs in Mathematics,
https://doi.org/10.1007/978-3-030-34732-1

19. W. Hensgen, An example concerning the Yosida-Hewitt decomposition of finitely additive measures. Proc. Amer. Math. Soc. **121**, 641–642 (1994)
20. T.H. Hildebrandt, On bounded functional operations. Trans. Am. Math. Soc. **36**, 868–875 (1934)
21. J.L. Kelley, *General Topology* (Van Nostrand Reinhold, Toronto, 1955)
22. R.B. Kirk, Sets which split families of measurable sets. Amer. Math. Monthly **79**, 884–886 (1972)
23. P.D. Lax, *Functional Analysis* (Wiley, Hoboken, 2002)
24. G.G. Lorentz, A contribution to the theory of divergent sequences. Acta. Math. **80**, 167–190 (1948)
25. J.C. Oxtoby Review: K.P.S. Bhaskara Rao, M. Bhaskara Rao, Theory of charges, a study of finitely additive measures. Bull. Am. Math. Soc. **11**(1), 221–223 (1984)
26. J. Rainwater, Weak convergence of bounded sequences. Proc. Amer. Math. Soc. **14**(6), 999 (1963)
27. W. Rudin, *Principles of Mathematical Analysis*, 3rd edn. (McGraw-Hill, New York, 1976)
28. W. Rudin, *Real and Complex Analysis*, 3rd edn. (McGraw-Hill, New York, 1986)
29. E. Shargorodsky. Personal Communication
30. W. Sierpiński, Sur les fonctions d'ensemble additives et continues. Fund. Math. **3**(1), 240–246 (1922)
31. L. Sucheston, Banach limits. Amer. Math. Monthly **74**, 308–311 (1967)
32. T. Tao, Amenability, the ping-pong lemma, and the Banach-Tarski paradox, https://terrytao.wordpress.com/2009/01/08/245b-notes-2-amenability-the-ping-pong-lemma-and-the-banach-tarski-paradox-optional/#more-1354
33. M. Valadier, Une singulière forme linéaire sur L^∞. Sém. d'Analyse Convexe Montpelier **17**(4) (1987)
34. A.J. Wrobel, A sufficient condition for a singular functional on $L^\infty[0, 1]$ to be represented on $C[0, 1]$ by a singular measure. Indag. Math. **29**(2), 746–751 (2018)
35. K. Yosida, E. Hewitt, Finitely additive measures. Trans. Amer. Math. Soc. **72**, 46–66 (1952)

Index

Symbols

(X, ϱ) a topological space, 9

$(X_\infty, \varrho_\infty)$ one-point compactification, 18

$((\mathfrak{G}, \tau)$, 57

$A_\alpha(u)$, 70

B^*, 22

$C(\mathfrak{G}, \tau)$, 3

$C^*(\mathfrak{G}, \tau)$, $C(\mathfrak{G}, \tau)^*$, 87

$C_0(X, \varrho)$, 18

$C_D(f, z_0)$, 84

$E \Delta F$, 14

G_δ

    closed, 24

    set, 24

    space, 24

G_δ-space, 25

$H^\infty(\mathbb{D})$, 84

L_p-spaces, 17

$L_p(X, \mathcal{L}, \lambda)$, 17

X_∞, one-point compactification, 2

$L_\infty(X, \mathcal{L}, \lambda)$, 18

$L_\infty(X, \mathcal{L}, \lambda)^*$, 27

$L_\infty^*(X, \mathcal{L}, \lambda)$, 3, 27, 39, 44

$\Pi(\mathcal{L})$ purely finitely additive measures, 36

$\Sigma(\mathcal{L})$, 13, 31

ba$(\mathcal{L})$ finitely additive measures, 31

ba$(\mathcal{L}) = \Sigma(\mathcal{L}) \oplus \Pi(\mathcal{L})$, 38

χ_A characteristic function, 28

ℓ_∞, 45

$\ell_\infty(\mathbb{N})$, 21, 45, 46, 71, 83

(†), v, 3, 16, 31, 42

$\hat{\mathcal{L}}$, 14

λ-finite intersection property, 44

$\mathfrak{G}$ 0-1 finitely additive measures, 3, 41

    Hausdorff topology, 57

    weak* closed in $L_\infty^*(X, \mathcal{L}, \lambda)$, 64

$\mathfrak{G}(x_0)$: $\mathfrak{G}$ localised at $x_0 \in X_\infty$, 78

$\mathfrak{U}$ ultrafilters, 3

$\mathbb{D}$, 84

$\mathbb{T}$, 84

$\mathfrak{D}$ Dirac measures, 5, 66

$\mathfrak{U}$, 42

$\nu_1 \leqslant \nu_2$, partial ordering, 33

$\nu^+, \nu^-, |\nu|$, 34

$\nu_1 \ll \nu_2$, absolute continuity, 35

$\nu_1 \perp \nu_2$, 35

$\nu_1 \vee \nu_2$, $\nu_1 \wedge \nu_2$, sup and inf, 33

$\overline{\mathcal{M}}$, 9

ω-almost everywhere, 42

$\omega_\mathcal{U}$, $\mathcal{U}_\omega$, 43

$\partial(E, F)$, 14

$\prec$, 7

$\preceq$, 7

$\mathcal{B}$, 9

$\mathcal{L}$, 8

$\mathcal{M}$, 8

$\mathcal{N}$ null sets, 10

$\mathcal{T}$, 57

$\mathcal{W}$, 63

σ-additive measures, 31

σ-additivity, 10

σ-algebra, 8

σ-compact, 18

$\sim$, 14

ϱ collection of open sets, 9

$\rightharpoonup$, weak convergence, 1

$\wp(S)$ the power set of S, 7

$\overset{*}{\rightharpoonup}$, 22

x_∞, point at infinity in $(X_\infty, \varrho_\infty)$, 18

(V), 1

(W), 1

© The Author(s), under exclusive license to Springer Nature Switzerland AG 2020

J. Toland, *The Dual of $L_\infty(X, \mathcal{L}, \lambda)$, Finitely Additive Measures and Weak Convergence*, SpringerBriefs in Mathematics,

https://doi.org/10.1007/978-3-030-34732-1

A
Absolute continuity, 35
Algebra, 8
Algebras and σ-algebras of sets, 8
Atom, 11
Axiom of choice, 8

B
Banach limit, 21, 53, 84
Banach space, 19
Base, 23
Blaschke product, 86
Borel
    σ-algebra, 9, 77
    measure, 77
    sets, 9

C
Canonical decomposition, 38
Characteristic function, 10, 14
Closed G_δ-set, 62, 79
Cluster set, 84
Cluster value, 77
Completely separable, 24, 60, 81
Complex function theory, 84
Convergence
    in measure, 10
    λ-almost everywhere, 10
Countable, 7
Counting measure on $\mathbb{N}$, 11

D
Darboux property, 11
Denumerable, 7
Dirac measure, 11
Directional limit of u at x_0, 82
Discrete topology, 19
Dual space, 1, 19

E
Equality almost everywhere, 42
Equivalence
    class, 8
    relation, 8
Essential range, 52, 77
    localised, 81
Extended real numbers, 7
Extreme point, 22, 64

F
Filter, 42
Fine structure at x_0 of $u \in L_\infty(X, \mathcal{B}, \lambda)$, 81
Finitely additive measures, 31
First axiom of countability, 23
Functionals, bounded, 19
Function spaces, 17

H
Hahn decomposition, 13
Hardy spaces, complex, 84
Hausdorff topology, 24

I
Inner function, 84
Integration, 47
    with respect to $\nu \in \mathfrak{G}$, 51
Isolated, 60
Isometric isomorphism, 58

J
Jordan decomposition, 13

L
Lebesgue
    σ-algebra, 12
    decomposition, 17
    integrable, 16
    integral, 15
    measure, 12
    point, 69
Lemma
    Urysohn, 24, 90
    Zorn, 3, 7, 8, 42–44
Local base, 23
Localising
    $\mathfrak{G}$, 77
    complex function theory, 84
    essential range, 81
    weak convergence, 80
Locally compact, 18, 77

M
Maximal, 8, 42
Maximal element, 8
Maximum norm, 18
Measurable
    functions, 9
    sets, 9

space, 9
Measure, 10
    σ-finite, 10
    Borel outer, inner regular, 11
    finite, 10
    Lebesgue, 12
    space, 10
Measure space
    complete, 10

N
Normal topology, 24
Null sets, 10

O
One-point compactification, 18
Oscillatory functions, 74

P
Partial ordering, 33
Poisson kernel, 85
Positive and negative parts, 13
Purely finitely additive measures, 36

R
Real measure, 12
Reconciling representations, 87
Regular
    Borel measure, 11
    inner, 11
    outer, 11
    real Borel measure, 13
Regular topology, 24

S
Second axiom of countability, 24
Separable, 22, 23
Sequential weak continuity, 67
Shargorodsky, 5, 84
Shift operator, 53
Simple function, 10, 73
Singleton, 7
Singular, 35
Splitting measurable sets, 14
Step functions, 73
Sub-base, 22, 23, 57, 63
Symmetric difference, 14

T
Theorem
    Alaoglu, 22, 64, 68
    Alexandroff, 91
    Baire's category, 3, 14, 15
    Banach–Steinhaus, 21
    Dini, 71
    Dominated convergence, 1
    Hahn–Banach, 5, 20, 28, 44, 69
    Krein–Milman, 23
    Lebesgue–Radon–Nikodym, 16, 90
    Lusin, 19
    Mazur, 21, 69
    Monotone convergence, 49
    Nikodym's convergence, 32
    Radon–Nikodym, 17, 50
    Rainwater, 23
    Riesz representation, 1, 20, 27, 87
    Stone–Weierstrass, 21, 59
    Valadier–Hensgen, 3, 5, 54, 56
    Weierstrass approximation, 21
    Yosida–Hewitt representation, v, 3, 27,
        50, 87
Totally disconnected, 60
Total ordering, 7
Total variation, 13
Translations, 74

U
Ultrafilter, 42, 43
Upper bound, 8

W
Weak*
    closed convex hull, 23
    convergence, 22
    sequential compactness, 22
    topology, 22
    topology on $L_\infty^*(X, \mathcal{L}, \lambda)$, 63
Weak convergence
    at a point, 77, 80
    criterion, 70
    of sequences, 21, 67

Y
Yosida–Hewitt decomposition, 89

Printed in the United States
By Bookmasters